Environmental Management

Environmental Management

Issues and Solutions

Edited by

Michael Atchia

and

Shawna Tropp

Published on behalf of
United Nations Environment Programme (UNEP)

by

JOHN WILEY & SONS
Chichester · New York · Brisbane · Toronto · Singapore

Copyright © 1995 by UNEP
Published in 1995 by John Wiley & Sons Ltd,
Baffins Lane, Chichester,
West Sussex PO19 1UD, England

Telephone National Chichester (01243) 779777
International (+44) (1243) 779777

Reprinted November 1995

Other Wiley Editorial Offices

John Wiley & Sons, Inc., 605 Third Avenue,
New York, NY 10158-0012, USA

Jacaranda Wiley Ltd, 33 Park Road, Milton,
Queensland 4064, Australia

John Wiley & Sons (Canada) Ltd, 22 Worcester Road,
Rexdale, Ontario M9W 1L1, Canada

John Wiley & Sons (SEA) Pte Ltd, 37 Jalan Pemimpin #05-04,
Block B, Union Industrial Building, Singapore 2057

Library of Congress Cataloging-in-Publication Data

Environmental management: issues and solutions / edited by Michael Atchia and Shawna Tropp.
   p.   cm.
   "Published on behalf of United Nations Environment Programme (UNEP)"
   Includes bibliographical references and index.
   ISBN 0-471-95518-3
   1. Sustainable development. 2. Environmental policy. I. Atchia, Michael. II. Tropp, Shawna.
   III. United Nations Environment Programme.
   HC79.E5E57843 1995
   363.7—dc20
                                                                94-35988
                                                                CIP

British Library Cataloguing in Publication Data

A catalogue record for this book is available from the British Library

ISBN 0-471-95518-3

Typeset in 10/12pt Times by Photo-graphics, Honiton, Devon
Printed and bound in Great Britain by Antony Rowe Ltd, Chippenham, Wiltshire

Contents

Abbreviations

AMCEN	African Ministerial Conference on the Environment
CITES	Convention on International Trade in Endangered Species of Wild Fauna and Flora
CMS	Convention on Migratory Species of Wild Animals
ECG	Ecosystems Conservation Group
EE	Environmental Education
EMINWA	Environmentally sound management of inland waters
FAO	Food and Agriculture Organization of the United Nations
GATT	General Agreement on Tariffs and Trade
GEF	Global Environmental Facility
GEMS	Global Environmental Monitoring System (UNEP)
GIS	Geographic Information System (UNEP)
GRID	Global Resource Information Database (UNEP)
IAEA	International Atomic Energy Agency
IAWGD	Inter-Agency Working Group on Desertification
ICRAF	International Council for Research in Agroforestry
ICSU	International Council of Scientific Unions
IEEP	International Environmental Education Programme
ILO	International Labour Organization
IMO	International Maritime Organization
INFOTERRA	International Environmental Information System
IPCC	Intergovernmental Panel on Climate Change
IPCS	International Programme on Chemical Safety
IRPTC	International Register of Potentially Toxic Chemicals
IUCN	World Conservation Union
MAB	Man and the Biosphere Programme (UNESCO)
NGO	Non-governmental organization
OECD	Organization for Economic Cooperation and Development
PEEM	Panel of Experts on Environmental Management for Vector Control
PIC	Prior Informed Consent
ROPME	Regional Organization for the Protection of the Marine Environment
SADCC	Southern African Development Coordination Conference

SCOPE	Scientific Council on Problems of the Environment (ICSU)
SDE	Sustainable Development Education
TFAP	Tropical Forestry Action Plan
UNCOD	United Nations Conference on Desertification
UNCTAD	United Nations Conference on Trade and Development
UNDP	United Nations Development Programme
UNEP	United Nations Environment Programme
UNESCO	United Nations Educational Scientific and Cultural Organization
UNIDO	United Nations Industrial Development Organization
UNSO	United Nations Sudano–Sahelian Office
WCIP	World Climate Impact Studies Programme
WCS	World Conservation Strategy
WHO	World Health Organization
WMO	World Meteorological Organization
WRI	World Resources Institute
WWF	Worldwide Fund for Nature

Foreword

This resource book has grown out of a series of seminars on environment management (called EMS) organized yearly from 1991 at UNEP Headquarters in Nairobi by the UNEP Environmental Education and Training Unit with inputs from, among others, the UNESCO Regional Office for Science and Technology in Africa. The participants, many of whom submitted the country case studies included in this collection, were executives or key members in a ministry or parastatal body concerned with the environment from a wide range of developing countries. This volume is a compilation of presentations made at these seminars together with the perceptions of participants and international civil servants from UNEP in the four following areas:

- the intellectual foundations of sustainable development;
- basic principles and processes in ecosystems;
- basic environmental management tools, including programme prototypes and funding possibilities; and
- specialized themes and strategies for environmental management.

Like all books that deal with scientific concepts and activities, this one will sooner or later be overtaken by time. Its contributors and editors none the less hope that a broad spectrum of its fundamental propositions and suggested strategies will endure.

During these month-long seminars, notes and records were kept on the mostly verbal presentations and enthusiastic discussions that followed. From over two thousand pages of material so recorded, the editors have produced this concise reader and therefore bear the responsibility for any inaccuracy appearing under an author's name.

The data accompanying the issues and arguments raised during the seminars have been deliberately omitted from this volume. For such data the reader is referred to resource books such as *World Resources* (published yearly by WRI, UNEP and UNDP) or the *World Environment* (1972–1992), published by UNEP.

Working within the design provided and developed by Dr M. Atchia, Chief of our Education and Training Unit, Ms Shawna Tropp did the bulk of the transcribing from audio records to text. Her tremendous efforts must find recognition here. The results of her work were further edited into this final text. Thanks are also due to

Ulf Carlsson, a Junior Professional Officer from Sweden who helped in proof-reading the final text.

The present co-edition by John Wiley and UNEP of this resource book should be of interest to all those whose profession it is to plan and manage the environment at national level, members of international environment and development agencies and NGOs, as well as those teachers, researchers and students in institutes of higher learning concerned with any aspect of sustainable development.

Reuben Olembo
Deputy Executive Director, UNEP

Preface

Education is the key to changing long-established patterns of social behaviour. It can help combat the unsustainable production and consumption patterns that are responsible for environmental degradation, loss of biodiversity, population growth beyond the capacity of systems, and unplanned urbanization. Training, especially in environmental management for sustainable development, can provide effective support (in the form of trained executives and technicians) to the political will of countries and communities for an improved and sustained quality of life for all their citizens. Hence the need for environmental management as a discipline.

This volume and the seminars on environmental management on which it is based launched a new process of interaction at UNEP: an open exchange of views with working environmentalists from developing countries. Consequently, almost half this collection consists of case studies and follow-up discussions with the participants. The book should be viewed as the distillation of a dialogue rich in challenging ideas and shared experiences.

Environmental concern became manifest in the late 1950s in developed countries as more and more citizens voiced alarm at the deteriorating quality of their air and water. Public outcries spurred government action on health hazards emanating from pollution. The role of the public at that time (and this cannot be underestimated) was crucial. This implies that governments may only succeed in tackling environmental issues by seeking support from their citizens and involving them in all actions undertaken to improve environmental quality. The rich crop of writings on the conservation of nature from the 17th to the early 20th century were useful to these early "green" leaders in their quest for a counterpart to industrialization and urbanization.

While the rich countries focused increasingly on the deterioration of the environment, the poorer nations perceived this concern as largely aesthetic. As late as 1972, many officials in developing countries opposed environmental programmes as attempts by the industrialized nations to impose a brake on their economic development. The Stockholm Conference began to examine the relationship between environment and development, a continuing debate that led to the 1992 World Conference on Environment and Development in Rio and the adoption by the world community of an agenda for the 21st century. This debate is far from finished. Each new surge of the human spirit will reveal a new relationship between environment and the process of development. Unfortunately, too, each new degradation of nature will require an adjustment between people and their surroundings, often global as well as local. The depletion of the ozone layer is a case in point.

Some ten years after Stockholm, a major shift took place in environmental thinking. We moved from the concept of preserving natural resources to that of the rational exploitation and development of those resources. The growing populations of developing countries require economic resources that can only be provided by economic growth. This growth, in turn, must be supported by a process of sustainable development that cannot be realized unless adequate natural, human and financial resources are available. In all cases, staying within the limits of the carrying capacity of both individual areas and regions and the planet Earth as a whole is at last being perceived as a necessity.

In developing countries, modern scientific and technical skills are scarce. The misuse of natural resources is still widespread, evidenced by the mismanagement of land, the degradation of soils, the pollution of air and water and the progressive destruction of forests. The current constraints on financial resources are well known, notably external debt, plummeting commodity prices and distorted exchange rates. The entire development process—including the crucial support of education, training, public information and the building of public awareness—falters. The result is diminished economic growth, lower per capita income, lower per capita expenditure and hence poverty.

To arrest and reverse this process, it is necessary to strengthen the management of natural resources, a key factor in economic development. The process clearly begins with monitoring the environment—determining its state and ascertaining what is available. Planning approaches will be applied and tested; adjustments will be made accordingly.

However, the way in which the components of the environment are dealt with is still far from satisfactory. They continue to be regarded as free goods that can be polluted or misused without direct cost. Environmental management must involve environmental accounting. This means asking the cost of a cubic metre of clean air as opposed to a cubic metre of polluted air. Far more complex, how does one calculate the cost of one hectare of tropical forest? Is it the cost of the wood, the land, the biological diversity it contains, the shelter it provides to indigenous peoples, its functions as a watershed or carbon sink, the value of the oxygen its plants produce? We need a fundamental definition of the cost of environmental action compared to the cost of inaction. Only when we formulate this within a cost-benefit analysis approach that rejects pollution as an externality can we determine the net cost of managing the environment properly.

Very few countries can handle by themselves the total impact of environmental degradation. Actions at the subregional, regional and global levels may well be far more effective. For example, control of marine pollution in the Mediterranean required a collective effort by all the Mediterranean basin nations. Similarly, the Gulf States have had to tackle the pollution choking their long, narrow, all-but-inland sea jointly. Such problems have led to the elaboration and adoption of binding international instruments. Further, it is now widely recognized that such dilemmas as ozone depletion, climate change and the transboundary movement of hazardous wastes have planetary implications. Consequently, global treaties have been and continue to be negotiated with the increasing participation of both developed and developing countries. The environment, an issue that only 20 years ago divided North and South, may yet unite us. The United Nations Environment Programme (with its headquarters in a developing country, Kenya) is instrumental in bringing about such international policies, agreements and actions. UNEP's role has been further enhanced by the UNCED process.

The mandate of UNEP, first developed in 1972, was written with foresight. Governing Council documents over the years have elaborated upon and made its elements more explicit. It has allowed the organization to build awareness of environmental issues around the world and to catalyse action. Of particular note has been the facilitation of the development of international environmental law, the catalysis of capacity building in developing countries and the introduction of environmental considerations into the social and economic policies, projects and programmes of UN agencies and world governments.

The United Nations Conference on Environment and Development exposed for world view the inevitable future of our current path of development. For the sake of the environment and for people a new way had to be found—that new path was sustainable development. World leaders adopted a common commitment to change through a statement of principles, an ambitious plan of action—Agenda 21.

In acknowledging the unique contribution of UNEP, UNCED offered a challenge to the organization. New questions were posed:

- How are environment and development linked?
- What are the ecological ground rules for sustainability?
- What is UNEP's particular contribution to the implementation of sustainable development?
- How can UNEP work effectively with a wide range of partners without duplicating their efforts?
- How can UNEP set priorities among a growing list of expectations and given the serious constraint of unpredictable and inadequate resources.

Agenda 21 reinforced UNEP's catalytic and coordinating role and assigned new or redefined responsibilities to the organization. One of UNEP's main currencies is information. Reliable flow of objective and meaningful information is critical to the mobilization of an environmentally aware and educated public and to the stimulation of effective decision-making.

Now (i.e. during the 94/95 biennium) and in an effort to sharpen our focus, UNEP's various programmes as reflected in this book have been clustered under three main headings in order to encourage integration and synergy and yield more effective results. These clusters are:

- international consensus building;

- national environmental management support, including national capacity building;
- environmental assessment, information for decision-making and disaster prevention, preparedness and response.

Under the leadership of the new Executive Director of UNEP, Ms Elizabeth Dowdeswell, these clusters and UNEP itself will surely evolve as we learn to be more responsive to the needs of countries, within the context of what (ecologically) can or cannot be done on this planet.

Michael Atchia
Programme Manager, Education and Training, UNEP

Part I

SOME FOUNDATIONS OF AND POLICIES FOR SUSTAINABLE DEVELOPMENT

Chapter One

Basic principles and concepts essential for achieving a functional environmental attitude

MICHAEL ATCHIA
Chief, Environmental Education and Training Unit, UNEP

THE CHALLENGE

Like many another educator, I take my texts from my surroundings and cull questions from them, as below:

> As we continue to waste the natural resources of our planet, new tensions are being created which, unless we as a family of nations mend our ways, will imperil the political stability of the whole world.
>
> The great danger is that we will not see the environmental dimensions behind the new conflicts. And we will seek to resolve disputes through arms—a reaction with a millennial pedigree.
>
> Mostafa K. Tolba, Executive Director, UNEP, speaking to the Royal Institute of
> International Affairs, London, 15 March 1990

Do we recognize fully the environmental dimension of individual and collective security?

> We found everywhere deep public concern for the environment, concern that has led not just to protests but often to changed behaviour. *The challenge* is to ensure that these new values are more adequately reflected in the principles and operations of political and economic structures.
>
> The Report of the World Commission on Environment and Development, 1987,
> p. 6

What do we need to modify in our present structures and actions in order to start meeting this challenge?

> Poverty itself pollutes the environment, creating environmental stress in a different way. Those who are poor and hungry will often destroy their immediate environment in order to survive: they will cut down forests; their livestock will overgraze grass-

lands; they will overuse marginal land; and in growing numbers they will crowd into congested cities. The cumulative effect of these changes is so far-reaching as to make poverty itself a major global scourge.

Ibid, p. 8

How far do you concur with this view of poverty? What, in your personal field of expertise and action, can you contribute to a sustainable solution to this problem?

For the wellbeing of the human environment, the 1990s must be the decade of decision. If we put off the hard decisions, it is not difficult to imagine what the world will be like 20 years from now: a world stripped of tropical forest and with a million or so fewer wild species; a world overtaken by sea-level rise and climatic disruption; a world of fierce competition for depleted resources.

Mostafa K. Tolba, Executive Director, UNEP, addressing the Economic and Social Council, 13 July 1989

Any action plan for the FUTURE which does not involve you, the children and youth of today, is doomed, for the future belongs to you.

An educator's creed, M. Atchia, at a Youth Round Table Meeting, 1988

Can all the remarkable improvements made this decade be of any use in the long term unless transmitted in a systematic way to future generations?

Thus, what knowledge, skills, and attitudes about the environment would you wish for in four-, eight-, 12-, 16-year-old children: 20-, 35-, 55-, 70-year-old adults?

It is my belief that *education* is the answer to the above challenge. For it is through education—practically oriented education—that children and youth obtain the awareness, the knowledge and the skills and acquire the attitudes necessary to pursue successful action for sustainable, environmentally sound development (or to put it in the terms utilized by a recent Council of Europe meeting, "development at acceptable conditions").

Furthermore, it is *science* and *science education* (and in particular, biology, as the science of life) that give us the conceptual framework for practically managing the *biosphere* (forests, soils, air, water, chemicals, energy, biological diversity and human beings themselves), as well as the technosphere (human habitats, industry, transportation, etc.).

How could one, for example, successfully promote and manage wildlife areas or agriculture, fisheries or human health and quality of life without a deep understanding of the ecological and physiological principles involved?

How could one, for example, successfully promote the development and use of clean technologies in industry without a deep understanding of the physico-chemical principles involved?

NEW GOALS AND DEFINITIONS

Environmental Education (EE) is currently assuming a connotation markedly different from the EE of the 1970s which was generally understood as school-based education *in, about* and *for* the environment. EE of the nineties is showing signs of becoming distinctly more scientific and technical than before with the strongest possible accent on direct application. Its main characteristics can at the onset be boldly defined through two intimately linked goals: education towards protection and enhancement of the environment; education as an instrument of development for improving the quality of life of human communities.

Protection and development taken together are now universally described, since the report of the World Commission on Environment and Development, as sustainable development, namely "development that meets the needs of the present without compromising the ability of future generations to meet their own needs" but more precisely the development of a capacity to convert a constant level of natural resources into increased satisfaction of human needs.

An environmental policy is a statement by a government of what it intends to do with, for or about the environment. Environmental programmes are a series of activities for monitoring and managing the environment (or supporting these activities) aimed at realizing an environmental policy. Most governments (heads of state or ministers in charge of such sections as natural resources, urban and rural planning, mining, agriculture and fisheries, industry, wildlife or the entire environment itself) have probably given thought to (if not actually defined) their environment policy in close conjunction with accepted national development policies. The main goal for this combined environment and development concern sounds, in a nutshell, like this: "achieving environmentally sound and sustainable development". Hence EE for the 1990s must become Sustainable Development Education (SDE) in support of the goal defined above.

The four major components of SDE advocated here and implied throughout this chapter are: the teaching and learning of the theory and practice of:

- Environmental *monitoring*;
- Environmental *protection*;
- Environmental resources *development* and *management*, and
- Environmental *enhancement*.

Referring to the World Commission's definition of sustainability, we educators have one remarkable advantage over everyone else: politicians, scientists, technologists and planners included. The future generations, at least those of the immediate future, are, so to speak, in our hands. Consider, for example, an imaginary girl aged six whom we shall call Victoria and who, in the year 2040 at the age of 52, will become the president of the United States. How grateful would those in the '40s and '50s of the next century be if, taking no chances, we took all of today's six-year-old Americans by the hand and taught them these elements of positive environmental action based on a respect for nature and a devotion to human well-being. And what is true for the future president of the United States is true for all other citizens of all other countries.

THE GOAL OF APPLICABILITY

The only justification for the development of SDE throughout educational systems in all countries is its applicability to our twin goals of protection and enhancement of the environment and development for improving the quality of community life—in other words, learning "how to do it" through acquiring skills appropriate to each age, situation and location. To be concrete let us take one example: biological diversity and the "why" and "how" of conservation. Since the publication of the World Conservation Strategy (1980) by IUCN, WWF and UNEP, "why conserve nature?" has been adequately dealt with at international, national and local levels. In most countries, such a theme is present throughout the non-formal and formal sectors of education. Thus, the case for conservation having gained universal acceptance from Kindergarten to golden oldies clubs, this notion only requires maintenance. *How to do it* is what we are after now. As applied to our example of biological diversity, this would mean learning in practice how to conserve the highest possible proportion of local, regional and global biological diversity (and the related habitats) while answering the multiple needs of human communities for these resources and habitats.

The scope of SDE in dealing with biological diversity would include the following:

- Knowledge and understanding of the main concepts of ecology;
- Survey and systematic inventory of biological diversities, their habitats and present utilization;
- Research into ecology and bio-geography of species and ecosystems;
- Research in assessment of potential benefits to be derived from rational use of these biological resources;
- Information and education of the general public and special groups (in particular the decision-makers) about the state and potential of "their" biological resources;
- Training in acquisition of technologies for collection, identification, protection, monitoring, use etc. of (biological) resources, including gene banks and depositories of live genetic resources;
- Training in management strategies and skills with regard to wildlife protected areas, endemic areas, area restoration and recovery and the further development of national conservation and development strategies;
- Periodic assessment of the state of local, national and global biological diversity, in particular endangered species, species over-exploitation, habitat destruction and fragmentation, effects of pollutants on organisms and ecosystems etc.;
- Recognition of the rights, knowledge and technologies of indigenous people, research into and incorporation of such traditional values and practices into national systems (e.g. in education, medicine, agriculture, conservation etc.);

- Transfer of conservation and biological resources-use technologies;
- Development of appropriate national and global legal instruments to protect and regulate use of biological diversity and in particular for rational conflict resolution over the use of biological diversity and, indeed, all other resources.

The scale of human impact on the environment having expanded so markedly over the past four decades (i.e. 1950–1990), in particular in regard to human population growth, a comparable expansion in our response is essential in order to maintain (or, in several cases, recapture) what Atchia (1978) described in the late '70s as the necessary "homoeostasis" between people and planet.

A keynote chapter is catalytic in nature, hence this single example (biological diversity) of how SDE is perceived will suffice. Curriculum developers will have no doubt noted the implications of this approach for educational programmes at all levels.

BASIC CONCEPTS FOR EE

Of special, indeed seminal, importance will be an agreement on the basic principles which must guide people and nations in their conduct towards each other and towards nature to ensure the future integrity and sustainability of planet Earth as a hospitable home for the human species and other forms of life.

Maurice Strong, Preparatory Committee for the United Nations Conference on Environment and Development, Nairobi, 6–31 August 1990

Any curriculum should be based on well-thought-out and clearly defined concepts that one wishes the learner to acquire. Several attempts have been made to define such a set of concepts for EE. Martin (1944) attempted to determine 300 major and 236 minor principles of the biological sciences of importance to general education. Caldwell (1955) developed 344 principles in earth sciences suitable for inclusion into the school programme. Hanselman (1964) developed 152 "statements" related to conservation, with an interdisciplinary approach in mind.

A major step in this area was taken by Roth (1969) who defined 112 concepts for environmental management. Atchia (1978) advanced a set of 40 essential concepts for EE, with the upper-primary and lower-secondary school students in mind; a selection of 24 of these concepts is given below as a sample of the theoretical basis for any curriculum development in environmental education. In other words what one would wish all children of the world to know, feel, and do about the environment.

Essential concepts for achieving positive environmental attitude

1. The size and range of population is regulated by available physical resources (e.g. space, water, air, food) and by biological factors (e.g. competition), hence the concept of *carrying capacity* of a natural environment.

2. The flow of energy and material through an ecosystem links all communities and organisms in a complex of chains and webs that invariably starts with plants.

3. The planet is made up of a number of interacting and interdependent components.

4. We live in "space-ship earth", a closed system characterized as having limits.

5. Human activities and technologies influence considerably the natural environment and may affect its capacity to sustain life, including human life.

6. A mode of life heavily dependent upon rapidly diminishing non-renewable energy sources (e.g. fossil fuel) is unstable.

7. The relations between humans and their environment are mediated by their culture (their mode of life and habits).

8. A clear difference exists between the natural needs of human beings and those wants and desires artificially created by advertising and social pressures.

9. Economic efficiency often fails to result in conservation of resources.

10. Rational utilization of a renewable source (e.g. rate of fish caught equal to rate of natural regeneration of fish population) is a sensible way of preserving the resource while obtaining maximum benefits from it.

11. Sound environmental management is beneficial to both man and environment.

12. Resources use demands long-term planning if we are to achieve truly sustainable development.

13. Elimination of wastes through recycling and the development of clean technologies is important to modern societies to help reduce the consumption of resources.

14. Certain artificial contaminants (e.g. radio-isotopes, mercury, DDT) are too long-lived or of such a nature that natural processes are unable to eliminate them readily.

15. The ocean has become the final dumping place for chemicals, oil, sewage, agricultural wastes etc. because these follow natural pathways from which they are released to reach the ocean.

16. Wildlife populations are important aesthetically, biologically (as gene pools), economically and in themselves.

17. Nature reserves and other protected wilderness areas are of value in protecting endangered species because they preserve their habitats.

18. Destruction of natural habitats by human beings is the single most important cause of extinction today.

19. Destruction of any wildlife may lead to a collapse of some food chains.

20. The survival of humanity is closely linked to the survival of wildlife, both being dependent on the same life-supporting systems.

21. The protection of soils and the maintenance of sustainable agriculture are essential factors in the survival of civilizations and settlements.

22. A vegetation cover (grass, forest) is important for the balance of nature and for the conservation of soil, besides being an exploitable natural resource.

23. Soil erosion is the irreversible loss of an essential resource (and must be prevented).

24. Cultural, historical and architectural heritage are as much in need of protection as is wildlife.

Meadows (1989), recognizing that environmental educators on every continent develop material and methods as varied as the different cultures and ecosystems on Earth, attempted to select and list eight key concepts which could underline all environmental education. These are cited here as further food for thought: levels of being, cycles, complex systems, population growth and carrying capacity, ecologically sustainable development, socially sustainable development, knowledge and uncertainty and sacredness, i.e. nature has its own value, regardless of its value to human beings.

Given different socio-cultural contexts and advances in the fields of education and in the sciences of the environment, it is my belief that one should determine, periodically, which concepts are needed and relevant (and in what order of priority) to a given society or region, for a given decade. This exercise should be undertaken now by all countries acting collectively by region, or individually, for the last decade of this century (i.e. 1990–1999).

THE AREAS OF CONCENTRATION AND THE WORK OF UNEP

As we are all (individuals, institutions, organizations or nations) finite and limited in what we can do in a given time, the idea of *areas of concentration* is appealing and realistic, especially in such a vast and all-embracing theme as that of environment. In 1989 UNEP defined eight areas above all others that we feel require concentrated effort over the next few years so as to achieve sustainable development and improve (if not maintain, in the face of increasing populations and mounting levels of environmental degradation) the quality of life of each one of the 5000 million people now on Earth. Most of these areas of concentration are biologically based and deal with the relationship of human beings to technology and nature.

Climate change and atmospheric pollution

The intent is to improve understanding of sources of atmospheric pollution and climate change, to identify policy options for minimization of the changes or adaptation to them; to work, if necessary, on legal instruments to address climatic modification and change in atmospheric conditions. Global warming and destruction of the ozone layer are the two problems most spoken of in this area.

Some specific issues

- Nowhere is interdependence of countries and regions on one another more apparent than in the interactions taking place in the atmosphere;
- Because of the "greenhouse effect" a global warming of the planet is predicted for the years to come; according to some studies, by the year 2030 the world's average surface temperature may rise by 1.5° to 4.5°C; the release of carbon dioxide from the burning of fossil fuel is probably the main cause of this warming;
- Such global warming could lead to a variety of environmental threats such as flooding of lowlying coastal areas, rise in water tables and further desertification;
- Depletion of the ozone layer and appearance in 1984 of a "hole" in the ozone over Antarctica; the destruction of ozone is caused by general pollutants released by industry, namely chlorofluorocarbons (CFCs) used as aerosol propellants and foams and in refrigeration;
- The depletion of the ozone layer would increase levels of ultra-violet radiation reaching the ground and this, in turn, would lead to more skin cancer, more eye diseases, and damage our immune systems; other environmental effects would be a decrease in the rate of photosynthesis and hence lower crop and timber yields.

Examples of UNEP's action

Obtaining the entry into force of a Global Convention (Vienna, 1985) and a Global Protocol (Montreal, 1988) for the protection of the ozone layer; in collaboration with several other organizations (e.g. WMO, IPCC), developing the Climate Change Convention which was signed at UNCED in June 1992.

Pollution and shortage of freshwater resources

The intent is to assist governments in the integrated, environmentally sound management of water resources as a crucial requirement of human welfare and sustainable development; clean and safe tap water for everyone by the year 2000 (or whichever date is realistic) is the sort of target aimed at in this area.

Some specific issues

- Lack of water is the greatest threat to life—human, animal and vegetal;
- Dirty polluted waters are the world's major cause of disease. More than a third of the world does not have safe drinking water. Out of a world population of 5000 million:
  - 1700 million lack safe drinking water,
  - 1200 million lack sanitation,
  - 1000 million cases of diarrhoeal disease occur annually,
  - 200 million cases of schistosomiasis occur annually while up to nine million people die annually as a result.
- More than 97 per cent of the water on Earth is sea water.
- Less than 1 per cent of the supply of fresh water is available for human use; the rest is locked away in glaciers and polar ice-caps.
- Industry uses less water than agriculture, but pollutes it to a far greater extent.
- The number of thirsty, dirty cities around the world is increasing; the number of garden towns decreasing.
- The International Drinking Water Supply and Sanitation Decade was launched by WHO in 1981 with a view to providing safe drinking water and sanitation for all by the year 1990. This has *not* happened.

Example of UNEP's action

In 1985, UNEP launched the Programme for the Environmentally Sound Management of Inland Waters (EMINWA), a major initiative aimed at safeguarding the world's fresh water, and is now working on other river or lake basins (e.g. Chad, Nile).

Deterioration of coastal areas and oceans

The intent is to protect, develop and enhance, primarily through intergovernmental cooperation, the coastal and marine environments upon which large numbers of people depend directly; at the same time, through international cooperation, to protect the open seas and their inhabitants (e.g. marine mammals).

Some specific issues

- Most of the wastes produced by industry, agriculture and homes end up in the seas, poisoning marine life.

- Some very sensitive and important marine organisms, such as coral reefs and algal plankton, are particularly affected by pollution.
- The oil industry, with its marine-drilling operations, transport of oil by tankers and use of oil-driven engines at sea, is contributing substantially to polluting the ocean.
- For centuries, marine life has provided a considerable proportion of human needs for food and other useful material; over-exploitation of certain species (whales, tuna, crustacea), coupled with pollution, is threatening the supply of food from the sea and the survival of many species, in particular certain whales and other marine mammals.
- The capacity of the oceans to absorb man-made chemical wastes has definite limits (though not as yet measured) like that of any natural system. Mass tourism to certain coastal areas and the prospecting and transport of oil are probably the major threats to oceans.

Example of UNEP's action

Over 120 countries take part in UNEP's Regional Seas Programme. Each of these programmes is an action plan for cooperation on research, monitoring, control of pollution and other hazards and development of coastal and marine resources. The Mediterranean Action Plan, whose implementation began in 1976, now has 17 signatories.

Land degradation, including desertification

Desertification remains unabated, with its menacing impacts on food production capacity and quality of life. UNEP has resolved to assist countries in the formulation and implementation of national soils policies and action plans to combat desertification, to reclaim desertified land for productive use and, above all, to maintain and improve the fertility of productive soils.

Some specific issues

- About 3500 million hectares (ha) of land—an area the size of North and South America combined—are affected by desertification.
- Every year about six million hectares of land are irretrievably lost to desertification, and a further 21 million hectares are so degraded that crop production becomes uneconomic.
- The rural population affected by serious desertification arose from 57 million people in 1977 to 135 million in 1984.
- The situation is likely to become extremely critical in the rainfed croplands by the year 2000, and to be little better elsewhere.
- Desertification is caused largely by human action—or lack of it; it follows that it can also be arrested by human action.

- Although more money is needed, the most successful attempts to control desertification have been through action which is low-cost, local, small in scale and run by those personally affected.
- The technical solutions, such as reforestation, improved farming techniques and better land use, are well known and have been successfully applied in many areas.
- In spite of this, the battle is being lost. A massive new effort to control desertification is required if declining productivity, erosion, famine and political chaos are to be avoided.

Examples of UNEP's action

UNEP's role is mainly catalytic and less of implementation. Its role in combating desertification, as in other areas, is to create awareness of the problem, help formulate strategy and coordinate United Nations' action;

UNEP was designated the agency to coordinate the United Nations Plan of Action to Combat Desertification which was formulated at the United Nations Conference on Desertification (UNCOD) in 1977. It did this through the Inter-Agency Working Group on Desertification and by working with other bodies, such as the Consultative Group for Desertification Control, to mobilize support for desertification projects;

Most activities are carried out in conjunction with other agencies, and programmes such as UNESCO, UNSO, FAO and WHO. Creating greater public awareness of the issues is a task to which UNEP itself has paid particular attention. It has organized press briefing guides on the subject and supported a number of documentary films and filmstrips. It was involved in the development of a convention to combat desertification, as recommended by UNCED, which was signed in October 1994.

Biological diversity

For ethical, social, economic, scientific and technical reasons, the conservation and utilization of biological diversity is more esential than ever for environmentally sound and sustainable development and the continued functioning of the biosphere and human survival. Emphasis is therefore being given to helping countries achieve the sustainable management of their forest lands, mainly in accordance with the Tropical Forestry Action Plan, and by designating areas for the conservation of biological diversity, as well as opening for signature and ratification an international convention on the subject.

Some specific issues

- Tropical forests cover 2970 million ha of the Earth's surface, but have been rapidly depleted over the past century.

- According to one estimate, at least 225 million more ha of tropical forest will be cleared or degraded by the end of the century; the rate of depletion is about 21 ha per minute.
- At present rates, nine developing countries will have exhausted their broadleaved forests within 25 years and a further 13 within 50 years.
- The main cause of deforestation is the need to expand agricultural land (though logging often leads indirectly to deforestation by opening up previously inaccessible areas).
- High though the cost of prevention may be—as is so often the case—it is trivial compared to that of continued inaction.
- The destruction and degradation of forests has widespread implications for human society, among them: supply of fuel, fibre, food, building material, shelter, medicine and a source of genetic material; furthermore, forests protect soils and watersheds, retain water and prevent floods and provide carbon sinks against atmospheric pollution.
- Forests are the homes of numerous indigenous peoples who often have invaluable knowledge of their forest environment.
- Although 33 developing countries are currently net exporters of forest products, only 10 are expected to be so by the year 2000.
- Deforestation also threatens the natural balance of upland watersheds and deprives the world of the genetic diversity on which it ultimately depends.
- Local attempts to protect and manage forests, as well as to establish new plantations, have been successful in many countries—though not yet on a significant global scale.
- Rural women, many of whose families depend heavily on forest or tree products, have proved powerful allies in conservation programmes; non-governmental organizations have proved most effective in implementing projects at the grassroots level.

Examples of UNEP's action

In 1980, UNEP, FAO and UNESCO began to prepare for a global plan of action for the wise management of tropical forests. Although they emphasized that such a plan would not infringe on nations' sovereign rights to use their forests as they saw fit, they failed to convince the countries that mattered: 21 nations sent representatives to a meeting in 1982, but these did not include major forest countries such as Brazil, Burma, Colombia and Zaire. The plan seems to have been premature and was subsequently abandoned.

The Tropical Forests Action Programme, developed in 1985, has five priority areas: fuelwood; forestry's role in land use; forest industrial development; conservation of tropical forests and institution strengthening. A set of Forest Principles was made part of the agreed outputs of UNCED.

UNEP among others helped develop the Biodiversity Convention that was signed at UNCED, in June 1992.

Hazardous wastes and toxic chemicals

The intent is to improve national capabilities for minimizing the impacts of hazardous wastes and toxic substances by dissemination of information obtained: by assessment and evaluation of toxic and hazardous substances, by promoting the integration of environmental considerations into industrial development, by supporting the preparation and implementation of international agreements and, above all, by promoting the development of low- and non-waste renewable technologies. Work is also needed on the transboundary movements of hazardous wastes, toxic and radioactive chemicals and on the prevention of waste dumping at sea, among other areas.

Some specific issues

- More than seven million chemicals are now known, and thousands of new ones are discovered every year;
- Many of the 80 000 or so chemicals in common use are potentially hazardous; some are in common use precisely because they are extremely toxic;
- Since the 1950s, chemical accidents have become more frequent and more serious;
- To mention only a few examples, during the past 10 years thousands were killed by toxic gases in Bhopal, low-level radiation swept over Europe and the Rhine was polluted by pesticides;
- The disposal of industrial wastes has also led to the serious pollution of ground water in many countries;
- Pollution of the Rhine and the Mississippi is so serious that neither may any longer be a suitable source of drinking water;
- Stricter controls at home may encourage manufacturers in developed countries to try to dispose of their wastes in developing countries, despite high transport costs;
- Africa taken as a whole is still relatively unpolluted but probably not for long. There is a need to identify what are the potentially dangerous industries and zones on that continent and take legal, political and technical action to achieve clean and sustainable industrial development.

Environmentally sound management of biotechnology

Biotechnology, in particular the advanced techniques of genetic engineering, could be of very considerable help to humanity, e.g. by obtaining agricultural plants that fix atmospheric nitrogen directly or resist droughts, pests or salinity and in the field of pollution control by bacterial action. It could also pose considerable threats to the environment and to the health and security of people. Hence the need for the sound management of this emerging area.

Protection of human health conditions and quality of life

The intent is to prevent the degradation of the environment affecting particularly the living and working conditions of poor people; to make people aware of the close relationship between the environment and their quality of life; and to educate them in ways of improving both.

These areas of concentration have been discussed and were in broad terms adopted in August 1990 in Nairobi by the Preparatory Committee of UNCED and form part of Agenda 21. *Caring for the World* (1991), published by IUCN, WWF and UNEP as a successor to the World Conservation Strategy (1980), emphasizes the interdependence of conservation and development; the need to maintain earth's stock of natural capital, and to live only on its income. It takes full account of economic and social as well as ecological requirements of sustainability. Action proposals in *Caring for the World* are based on the following imperatives:

- Limiting human impact on the biosphere to a level that is within its carrying capacity;
- Maintaining Earth's stock of biological wealth;
- Using non-renewable resources at rates that do not exceed the creation of renewable substitutes;
- Aiming for an equitable distribution of the benefits and costs of resource use and environmental management among and within countries;
- Promoting technologies that increase the benefits from a given stock of resources;
- Using economic policy to help maintain natural wealth;
- Adopting an anticipatory, cross-sectoral approach to decision-making;
- Promoting cultural values that help achieve sustainability.

SDE IN PRACTICE

This final section deals with some aspects of feasibility, since, as has been stated earlier, the key aspect of SDE is considered environmental action. It is generally recognized that projects of a voluntary, social nature succeed when they fulfil the following criteria:

- They satisfy needs and aspirations of the community concerned;
- They are easy to carry out, fall within the capability, instrumentation and financial strength of the external agents and community members;
- They show some concrete results within a short time span (typically a plantation project that involves sowing seeds showing a good crop of fresh seedlings after a few days);
- They involve many levels of the community from the planning stage throughout implementation;

- They are easily reproduced, preferably without external help (i.e. people can say, "We ourselves can do this", which is a typical reaction of a well-assimilated educational experience).

Sustainable Development Education should be like voluntary social work—starting from identifying the needs of a community through the process of developing a conceptual basis and culminating in practical environmental action. It would therefore fulfil the same criteria for success or failure.

How does one determine the social and environmental needs of people? There are a variety of methods, ranging from observation through questionnaires and survey on to in-depth studies. It suffices here to state one basic element: needs as perceived from within a community itself have a different and certainly 'more real' value (although sometimes less objective) than needs of a community as perceived or, more often, decided upon by external agents or central planners. The considerable amount of work done by the Commission of Education of the International Union of Biological Sciences has shown the inconstancy (both geographical and temporal) of human needs, within, however, a framework of universal principles of life, as well as the gap between what is externally believed to be the needs and problems of a community and the real needs (Atchia, 1982).

In 1985, I made a first attempt to determine the factors responsible for quality of life in different communities, through a survey conducted in six countries. The two questions asked were simple:

"What does the statement quality of life mean to you?"
"In order to have a high quality of life, I consider the following elements important . . ."

The results distinguish what I called "essential" needs from those which are "peripheral" and represent only a sample of what more in-depth research could reveal. The six factors which were identified as essential needs were: education, environment, work, money, health and social relationships. Such an exercise, using the same two questions, is suggested as a starting point for the determination of needs of a community leading to the eventual development of an educational package aimed at improving quality of life. I will call this exercise 'the sustainable development education response'.

The step after "determination of needs" is the selection of scientific and social concepts which are appropriate to the given community and its needs. A sample of 24 essential concepts was given above; these represent what I would wish all children of the world to *know*, *feel* and *do* about the environment. Each given community could use these 24 concepts in this or a modified form for a start and add others as appropriate.

The next step in developing an SDE response is to transform a concept into learning material which would culminate in concrete environmental action where one or more development skills would be acquired. Let us take, for example, Concept 20: "The survival of humanity is closely linked to the survival of wildlife, both being dependent on the same life-supporting systems". A tree as an ecosystem could teach us this concept, and the lesson would naturally end with tree planting. In dry areas, such as those in many parts of Africa, Asia and America, learning to plant and care for trees is an essential sustainable development skill.

From there, one can move to another concept, e.g. to Concept 10, "Rational utilization of a renewable source (e.g. rate of fish caught equal to rate of natural regeneration of fish population) is a sensible way of preserving the resource while obtaining maximum benefits from it" and discuss the level of exploitation of forests (or fish or any other renewable resource) together with the benefits and practical uses of the products (timber, plants, medicines, fuelwood, fish and other marine animals, algae etc.). The final step would again be the learner's development of a practical skill (learning to fish, to do woodwork, to cook in the wild, to recognize and use medicinal plants etc.).

The same procedure would be applicable to the remaining basic concepts.

REFERENCES

Atchia, M. (1978) *Concepts and dynamics of environmental education*, doctoral dissertation, University of Salford, UK.

Atchia, M. (ed.) (1982) "Research in community-based biological education", IUBS–CBE.

Atchia, M. (1988) "A Case for Developing an Environmental Curriculum Adapted to the Needs of the Learners", paper presented at International School Seminar, Nairobi.

Atchia, M. (1990) Biology and the Quality of Life, in Rex Meyer (ed.), *Selected Issues in Biological Education*, IUBS–CBE.

Caldwell, L. T. (1955) Determination of earth science principles, *Science Education*, **39** (3), 196–213.

Connor, J. V. (1990) *Environmental Education in a Developing World*, UNICEF, New York.

Hanselman, D. K. (1964) *Inter-departmental Teaching of Conservation at the Ohio State University*. The Natural Resources Institute.

IUCN, WWF, UNEP (1980) *World Conservation Strategy*, Gland, Switzerland.

IUCN, WWF, UNEP (1991) *Caring for the Earth*, Gland, Switzerland.

Martin, W. E. (1944) *A Determination of the Principles of the Biological Sciences of Importance for General Education*, Unpublished Ph.D. dissertation, University of Michigan, Ann Arbor.

Meadows, D. H. (1989) *Harvesting One Hundredfold*, UNEP, Nairobi.

Report of World Commission on Environment and Development (1987) *Our Common Future*. Oxford University Press, Oxford.

Roth, R. E. (1969) *Fundamental Concepts for Environmental Management Education* (K-16), doctoral dissertation, University of Wisconsin.

Schaefer, G. (1989) "Biology as an Integrating Science", paper presented at IUBS–CBE Meeting, Moscow.

UNCED, *Agenda 21. The Report of UNCED* (1992), UN, New York.

UNEP (1989) Fifteenth Session of the Governing Council of UNEP, *Proceedings*.

UNESCO–UNEP (1986) *Procedures for Developing an Environmental Education Curriculum*, EE Series No. 22, IEEP, Paris.

Chapter Two

Human ecology and environmental ethics

LIVINGSTONE DANGANA
Programme Specialist, Ecological Sciences, UNESCO/ROSTA

AND

SHAWNA TROPP
Consultant to UNEP, Nairobi

INTRODUCTION

At this point on the eve of the 21st century, it has become clear that humankind is facing one of the greatest challenges of recorded history: how to reconcile continued economic growth and all that this implies with the constraints imposed by a shrinking resource base and an increasingly degraded environment. If we do not want merely to cast ourselves on the seas of change, we should use our knowledge of the historical processes that helped shape human technological powers to orient ourselves. Human ecology is a useful tool towards this end. Human ecology is concerned with relationship to the global environment. This vast field includes the patterns of development of past and present human societies as they have sought, by regulation and adaptation, to come to grips with a changing world throughout geological and historical time.

Our discussion will focus first on the developmental needs and management issues that have concerned human beings throughout historical time, and then outline UNESCO's experiences and scientific contributions to the continued search for viable environmental management options. It will conclude with a note on environmental ethics as a force behind the current global efforts to rescue Earth and humankind from impending disaster.

HUMAN AND ENVIRONMENT AS EVOLUTIONARY PROCESSES

The human race and nature are products of evolutionary processes, subject to the final test of survival. And the evidence of human resourcefulness in the art of environmental husbandry can be charted throughout our evolution from Neolithic food-gathering bands to agro-industrial society. We shall therefore hereunder consider the relationship between the inner dynamics of beliefs, values and goals and the outer dynamics of agricultural, industrial and ecological processes—of human historical transformation—in order to discover how we have survived through the creative values of our inner evolution.

Evolutionary patterns

Patterns in the outer processes of human transformation

We have grown progressively in technological sophistication, innovations in tools to work the land, produce goods, increase crop yields, develop the means of transport and production, transmit and receive information. The industrial revolution introduced steam-driven tools and machines, automation and, far more recently, telecommunication cybernetics. Each shift from one stage to the next in human evolution has been marked by a progressive and radical increase in the production and use of energy: from the energy of the food we ate to the additional energy of fire; the energy of harnessing domesticated animals; energy from coal, water and wind power through energy to electricity and atomic energy. This evolution was also characterized by a progressive increase in the use and variety of materials. From the rock tools of Neolithic man, wooden tools developed with the invention of fire. Soft metals such as copper came next, then iron in about 1450 BC, steel in 950 AD and, more recently, the base materials for modern alloys. Progressively too, we extended the capacity of our environmental control. Whereas our Stone Age ancestors' environmental control was limited to their hunting territories, even simple horticulture expanded our grasp to encompass a village and its surroundings.

In the more advanced agricultural societies of the Nile, the Tigris and the Euphrates, the Indus valley and the Ganges plain, "our" territory was extended along the navigable rivers; irrigation systems extended human control over the obstacles posed by the natural environment. Later, the high civilizations of Mesopotamia, Pharaonic Egypt, Mohenjo Daro, Crete, ancient China, the Olmecs and Mayans in Middle America and classical Greece and Rome expanded environmental control over vast areas, on sea as well as land. Their citizens, subjects and slaves cut down forests, built vast road networks and created greater cities that enjoyed piped water, hot and cold. These high civilizations also developed phosphate and nitrate fertilizers and heavy, albeit clumsy agricultural machinery that made it possible to cultivate a variety of crops beyond the natural ecology of their respective environments. Although much of this ancient technology was later lost, during the European industrial revolution a lot was reinvented, expanded and refined. Currently, our reach extends into the air and under land and sea. Unfortunately, however, the natural environment has begun to suffer seriously from pollution belched and blown into the atmosphere, the water systems and the oceans, often from remote sources.

Increasing interactions among ourselves and between ourselves and the natural environment have marked our progressive march from relatively self-sufficient food gatherers to food-producing societies with surpluses to sell and time to devote to other activities; to the wide land and maritime trade of the great ancient civilizations, often ranging from continent to continent, especially among Asia, Africa and Europe to the far-flung colonial empires of the medieval Arabs, Persians, Indians and Chinese to those of the Portuguese, Spaniards, and the Dutch and later, the English, French, Germans and Belgians in the 19th century. Human global interaction has bred increasing interdependence on a planetary level during the last 500 years. Now, there are no longer any self-sufficient nations economically, politically or, most important, ecologically. Information flows during the last 10 years alone have so dramatically improved that the word of mouth of Stone Age days is transmitted round the globe the very moment it is spoken via electronic devices that have made the world a global village, however quarrelsome.

With all these advances, humans have found themselves faced increasingly by complex problems. Early communal bands with relatively simple divisions of labour gave place to theocratic polities in many ancient civilizations, all of which developed elaborate social hierarchies and administrative and legal structures. Whether our current nation states are more complex or better coordinated socially and economically is a question better left to history. Certainly, however, regional organizations have since evolved and multiplied; and world-wide non-governmental organizations and transnational corporations have further complicated the human picture. Certainly, the individual now has multiple roles and allegiances and very few people have a clear and adequate knowledge of anything save a fraction of the structures and roles of their society.

Patterns in the development of the inner dynamics of human societies

Emphasis in the evolution of our inner dynamics—our beliefs, values and goals—has shifted back and forth between humans and the cosmos they projected, as well as the individual and the collectivity. At the dawn of socio-cultural evolution, the food-gathering societies worshipped the world they perceived, together with the forces and spirits they attributed to it. Later food-producing societies added a new interest in the heavens. The sun and the celestial bodies as these affected the seasons and their livelihood became important subjects of study, however elementary. More advanced agricultural societies shifted their gaze fully from the Earth to the cosmos. As above, so below: human societies were stratified in their respective images of the celestial. The European classical civilizations were anthropocentric; thus man became the measure of all things—*but* in a harmonious relationship with the cosmos. Western Christianity asserted that man had been created in the image of God and pictured a great chain of being stretching from the Deity to lead, which was believed to be the lowest of the elements. But man was at the mid-point of this chain, chained also in large measure to the social station to which he had been born and the multiplicity of duties of that station, whether he was a serf or king. The European Renaissance and Enlightenment, however, shifted human attention to the values of personal freedom and radical artistic and scientific experimentation, a few thinkers even arguing that careers should be open to talent. Moreover, men were now sure that they could dominate nature and that they should exploit it. Previous visions of human/cosmic harmony did not altogether die throughout the Western world, but there arose a faith in science and technology that led to the belief that infinite economic growth was both desirable and feasible, *that what was technologically possible was also humanly beneficial* and *that the earth was inexhaustible.*

However, rapid changes during the last few decades have begun to raise serious doubts about these views. The dynamics of modern industrial society seem to be collapsing, perhaps because the two major streams of European economic philosophy, capitalism and communism, never really took into account ecological factors. Neither Adam Smith nor Karl Marx ever realized that Europe's great economic revival in the 17th century was due in no small measure to two ecological phenomena. First, the black rat drove out the brown rat, which was host to the flea that was the reactor of bubonic plague. Second, a slight warming of the south European Atlantic drove huge schools of herring northwards towards the Netherlands, becoming the base of Dutch wealth that drove their enormous commercial expansion. One should also remember that the Spanish brought back to Europe from South America their most valuable discovery—not gold or silver but the potato—which became a widespread new nutrients-rich basic European food, probably a major factor in the improved health and growth of Europe's population, the base of its industrial workforce.

Resourceful as we are, human beings are beginning to ask on a world-wide scale if a simple continuation of existing trends will not lead to a destabilization of the ecosystems, if we are not already suffering from an information overload, if the beneficence of agroindustrial technology as we now know it does have upper bounds—and if, short of colonizing other planets, population growth curves should not flatten out. All this is a consequence of the undercurrents of environmental ethics, welling out of our inmost dynamics, i.e., our sense of equilibrium within the world in which we find ourselves.

ENVIRONMENTAL ETHICS

A code of conduct

We have much evidence that not only our technologies but also our values, beliefs and goals have evolved—our inner powers for adaptive change and development. Despite the dominant trends of European thought at the time, since the 16th century many perceptive observers have cried out against the despoiling of the environment. Shakespeare, the most popular playwright of his day, was one of them. Later thinkers, such as Rousseau and Darwin, were others. Nor did they represent a small minority; all but illiterate farmers and factory-hands shared their views, if only because they saw how dirty their own air and water were becoming.

Such seems to be the force behind current trends within the United Nations and its member governments in their search for an agenda for global sustainable development, as manifest in the World Commission for Environment and Development (1987) and the Earth Charter of the United Nations Conference on Environment and Development (1992). This force derives from the development of an ecological psyche, which itself derives not only from our improved understanding of the ecological facts of life, but from all the world's great religions, which have always extolled nature's beauty and integrity, as well as laws and regulations concerning the environment that have existed for centuries and led to current national and international environmental laws and regulations. It is these positive tenets and codes of conduct that emphasize ideal human behaviour with respect to the environment that constitute environmental ethics. The human relationship, rights, obligations and responsibilities *vis-à-vis* the environment hold the key to our mutual survival with planet Earth.

The World Commission for Environment and Development (WCED)

The World Commission for Environment and Development was called upon by the General Assembly of the United Nations to draw up a global agenda for change in our patterns of environmental use. The clear message of the report, named after the chair of the Commission, Prime Minister Gro Harlem Brundtland of Norway, reminds all that we have only one Earth for our survival. The report touched the sensibilities of many people, on the imperative of sustainable development, but, more importantly, it paved the way for the heightened global concern that preceded the Earth Summit of 1992.

Scientific innovation is an indispensable step towards the search for sustainable development. UNESCO's Man and Biosphere Programme (MAB), launched in 1968, is one such endeavour which seeks to match science and ideas in the search for human survival on earth.

MAB AND HUMAN ECOLOGY

UNESCO's programmes on environment and natural resource management (notably MAB) aim at providing the scientific basis and trained personnel needed for solving the environmental problems of our times. Traditionally, MAB has involved research by interdisciplinary teams on ecological and social systems, field training and the application of a systems approach to understanding the relationships between natural and human components of development. Today, the overall goal of MAB is to help establish the scientific basis for sustainable development and to assist countries in the development of their human resources. Although there were 14 defined areas of research, the financial and human resources of MAB were concentrated more or less on the following six main areas for the first decade of the programme: coastal areas and islands, humid and sub-humid tropics, arid and semi-arid zones, temperate and cold zones, urban systems, biosphere reserves.

The biosphere reserve structure performs three major functions, which have evolved with the development of MAB and other work being carried out on biological diversity within the United Nations' system. Each of the more than 300 reserves in 175 countries worldwide:

- serves as a repository for the conservation of *in situ* representatives of Earth's main ecosystems;
- acts as a research site for monitoring these ecosystems;
- provides a laboratory for seeking ecologically sound solutions to the land use problems of the local people, involving them directly in that effort.

These three objectives are realized through a three-tier zoning system that can be adapted to a variety of geographic and economic situations. Each biosphere consists of one (or several) minimally distributed *core* areas, each of which protects a significant inland or coastal ecosystem. Around this core stretches a *buffer zone* devoted to basic and applied research, environmental monitoring, traditional land use, recreation and/or tourism or environmental education and training. The third, *outer zone*, is a transition area in which research is applied to local community needs. Thus, the system provides a basic framework for cooperative studies on biological diversity and monitoring change in natural and semi-natural ecosystems.

As a result, UNESCO has a wide range of pilot projects throughout the world: the building of research facilities in developing countries; the accumulation of impressive environmental data; and the forging of a new research tool—the international network of biosphere reserves. But MAB itself, an evolving system, has had to develop a change of perception regarding the human role within its programme. Whereas earlier studies focused on the characteristics and processes of the natural environment and the impact of human actions in disturbing a system that would otherwise be more or less in equilibrium, the growing trend in recent years has been to consider environmental and biosphere changes caused by human activities as an integral part of a continually changing and interacting environment. In other words the idea of Man *and* the Biosphere has evolved to one of Man *in* the Biosphere. This is why the new orientations of MAB research at the close of the century focus on intensities of human impact, human-impacted resources, human investment, and human responses, as shown by the following research themes:

Ecosystem functioning under different intensities of human impact

Better understanding of ecosystem functioning is necessary for three reasons:

- a certain type of use may be considered as being sustainable for a relatively short period depending on the socio-economic context;
- ecosystems may be affected by anthropogenetic factors coming from outside the system;
- ecosystems face short- and long-term natural changes imposed from both within and outside the system such as climatic changes. This proposed orientation

emphasizes comparative studies of the important processes in the functioning of ecosystems, subject to different intensities of human impact. Studies will include both regional comparisons of ecosystems affected by similar degrees of human impact, and local comparisons often based on biosphere reserves, of the relative importance of processes within an ecosystem that is subject to different degrees of human impact. The importance of a social historical perspective of system responses to human impact must be emphasized.

Management and restoration of human impacted resources

Ecosystem restoration is needed to recover the loss of amenities of ecosystems that have been heavily damaged. While under some circumstances the restored state may be a mature or climax ecosystem, this is not necessarily the case. Human beings have a great variety of needs that can be satisfied by an equally variable set of ecosystem states that can be considered as desirable under the proper circumstances. The research units addressed under MAB would not only be ecosystems but also resource management systems. Field research will take the form of manipulative experiments to learn about the susceptibility of human impacted ecosystems to management and research designed to develop management prescriptions for specific situations or management problems.

Human investments and resource use

Investments of knowledge, time and money drive change. Such investments can be destined for improving social welfare, for furthering economic development, for enhancing resource sustainability or for all three purposes. If the investments are narrowly based, however, the consequences can be the reverse of those desired. Those reverse consequences are not only local or regional; they are also global. Therefore the link between human welfare and ecosystem sustainability lies in investment processes that reflect local, regional and global forces. For this reason this orientation requires the integration of knowledge of social perceptions and expectations of the behaviour of biophysical systems and of the process of investment, disinvestment and reinvestment. This orientation emphasizes change and evolution, thus providing a particular opportunity to extend the procedures and methods used in the MAB programme.

Human response to environmental stress

There are many examples of environmental influences on human health and welfare such as the impacts of natural hazards, noise pollution, heavy metals, pesticides and fertilizers. Some human environmental changes have many positive impacts on health and these environments are suitable for sports activities and as a means for restoring the energy, health and general spirits of people. The MAB programme, with its strong combination of natural and socio-economic studies of human ecology, has the opportunity to make an important contribution to ongoing environmental health projects and assume a leadership role in such multi-disciplinary research dealing with positive and negative impacts of environmental changes on human health and welfare.

CONCLUSION

People are rapidly approaching a crossroads of crises. But drawing from our past and the new-found activities, driven by environmental ethics and scientific innovation within individual nations and the international community, there is some hope for man and planet Earth.

DISCUSSION

Question: Man is an insider in the environment. Is he a predator as well?
Answer: MAB grew out of fears of growing human predatory tendencies. For example, when human beings lived in forests, they didn't change those ecosystems more than any other animal until they moved outside and began to destroy forests—or parts of them—altogether.

Question: How much use does MAB make of local expertise in the biosphere reserves?
Answer: As much as possible; local people have often carefully guarded the secrets of their lands for centuries. A first step in the establishment of a biosphere reserve is winning local confidence. MAB experts, as well as the people of the area, need education.

Question: Since the pharmaceutical industry makes use of plants for medicinal purposes and since UNESCO conducts research on plants, shouldn't the pharmaceutical companies assist MAB research?
Answer: Ideally, yes, since we are speaking of preserving resources for future generations. We are working closely on this issue with WHO, as well as other United Nations' bodies.

Question: What are your suggestions on the transfer of technology to developing countries?
Answer: I have no specific suggestions. My point is that the environment is an area in which global cooperation is necessary, with emphasis focused on local participation for the solution of local problems.

MAJOR ACHIEVEMENTS

Some of the more notable and recent achievements of MAB include the following:

● MAB established an international network of biosphere reserves, which are protected areas representative of the world's major ecosystem types. The biosphere reserve concept is recognized as one of the most distinctive contributions to the promotion of participatory approaches to the conservation and sustainable use of biological diversity. The international network of biosphere reserves constitutes the only network of protected areas at the intergovernmental level. As of mid-1991, the network comprised 300 biosphere reserves in 75 countries, covering nearly 164 million hectares.

- MAB evolved, through the work in biosphere reserves and specialized projects, interdisciplinary, cross-sectoral approaches to sustainable development in different ecological and socio-cultural settings, including arid and semi-arid zones, Mediterranean systems, small islands, mountainous areas etc.
- MAB set up networks for comparative studies on topics which scientists around the world consider of vital importance (e.g. effects of land-use changes in Europe; studies on land/freshwater interfaces, tropical soil biology and maintenance of fertility; pastoralism in semi-arid zones).
- MAB promoted regional cooperation on natural resource issues through regional field projects (e.g. the Sahel) and technical seminars.
- MAB published technical syntheses, useful for scientists and resource managers, on such topics as the practical control and restoration of eutrophied lakes and rivers; environmental management of small islands; rain forest regeneration and management; reproductive ecology of tropical forest plants; and debt-for-nature swaps.
- MAB contributed, over the period 1969–1990, to the professional training of 1,200 specialists through more than 50 training courses and launched the MAB Young Scientists Research Awards, with 30 grants awarded by the MAB Bureau between 1989 and the end of 1990.
- MAB mobilized resources for building up the capacities of developing countries for integrated approaches to resource management, for example in China (with financial support from Germany) and in the Hindu Kush–Himalaya region, with the consolidation of the International Centre for Integrated Mountain Development (ICIMOD).
- MAB prepared a series of 36 posters—"Ecology in Action"—as an experiment in communicating scientific information to a broad public audience. UNESCO produced the materials in three languages, but member states have since translated them into 40 additional languages and have modified them for local use.

SELECTED BIOSPHERE RESERVES

1. *Charlevoix (Canada)* Strong support from the local community to conserve the natural and cultural heritage using the biosphere reserve concept. A "Cooperation for the Charlevoix World Biosphere Reserve" in particular promotes the awareness of local people to environmental questions within Quebec and elsewhere.

2. *Central California Coast (USA)* A complex area of land and sea around the San Francisco area. Consortium of universities heading research efforts. Was the site for a UNESCO–IUCN workshop in 1989 to develop guidelines for applying the biosphere reserve concept to cover the land/sea interface in coastal areas.

3. *Maya (Guatemala)* A 1 million ha complex in a highly diverse rain forest area declared a biosphere reserve under a special congressional decree. Contains many areas of cultural significance such as Tikal National Park. Interest in sustainably harvesting rain forest products.

4. *Tijuca–Tingua–Orgâis (Brazil) (Rio de Janeiro), Serra da Ribeira–Valle da Graciosa (Brazil)* The first biosphere reserves in Brazil designated in 1991 concern two important remaining portions of the Atlantic Forest which has been reduced to 3 per cent of its original areas. The focus of research and environmental education campaign especially for the urban population of Rio de Janeiro.

5. *Mananara Nord (Madagascar)* The first biosphere reserve in Madagascar, protected by a special decree and the site of a UNESCO–UNDP integrated development scheme to provide improved food, health and educational opportunities to the local community and thereby to save an intact tract of tropical forest harbouring many rare and endemic species such as the recently re-discovered hairy-eared dwarf lemur (*Allecebus trichotis*).

6. *Tassili N'Ajjer (Algeria)* Over 7 million ha of the Sahara Desert famous for its important rock painting depicting animals which lived there in prehistoric times when the climate was much wetter. A UNESCO–UNDP project proposes reconciling heritage conservation and the needs of the sedentary and nomadic tribes in the face of increasing tourism.

7. *Mare aux Hippopotames (Burkina Faso)* An area of typical West African savanna and woodland with a local population of over 22,000 people, this site benefited in 1989 from a study tour organized by the Ecole Nationale des Eaux et Forêts (EBGREF) of France which worked with national research and training institutions to prepare a revised management plan.

8. *Sinharaja (Sri Lanka)* The last patch of primary lowland rain forest in Sri Lanka. The management regime foresees the use of a certain palm, kitul, and several species of cane by local people in the buffer and transition zones. A buffer zone development project financed by NORAD is attempting to harmonize the different functions of the biosphere reserve.

9. *Palawan (Philippines)* Designation of a biosphere reserve in 1990 was facilitated by a feasibility study made by the MAB National Committee of the Philippines. Subsequently, a Strategic Environmental Plan for the Island of Palawan was drawn up with the support of the European Union. The biosphere

reserve includes all areas covered by the Ecologically Critical Area Network (ECAN) identified under this plan. The ECAN covers almost the entire island and some coastal areas.

10. *Changbaishan (China)* With large areas of untouched temperate forest, this is one of the research sites of the Cooperative Ecological Research Project jointly supported by the Chinese Academy of Sciences, Germany and UNESCO to advance knowledge on the conservation and use of forest systems in China.

11. *Fitzgerald (Australia)* In the Mediterranean climate area, this site conserves its many endemic plants and brings scientists and local farmers together to resolve the problems of salinization and erosion resulting from ill-adapted agricultural practices in the buffer zone.

12. *Cévennes (France) and Montseny (Spain)* These two biosphere reserves with similar ecological features have been "twinned" since 1987 with a cooperative programme of research and training. Activities include the joint preparation of brochures for the public, an exhibit on the conservation of the Mediterranean environment area and school exchanges.

13. *Rhôn (Germany)* Within the federal government system in Germany, this site is especially interesting as an example for regional planning since it lies in three separate states or *Länder*, one of which belonged to the former Democratic Republic. The aim is to coordinate research and improvements to land use for the benefit of the local people, and to monitor environmental changes.

14. *Oka River Valley (Russia)* Under a bilateral USSR–USA arrangement, there have been exchanges of scientists within biosphere reserves in both countries to develop a protocol on the long-term monitoring of pollutants using similar methods and to identify pairs of analogous sites for research on biological diversity. The Oka River was chosen for the visit of US scientists in 1989.

Chapter Three

"The environmental perspective to the year 2000 and beyond": how to make use of this policy statement

UTTAM DABHOLKAR
Deputy Coordinator, Support Measures, Office of\the Environment Programme, UNEP

THE NATURE AND CONTENT OF THE ENVIRONMENTAL PERSPECTIVE

Looking at the "Environmental Perspective to the Year 2000 and Beyond", many an environmental manager may well ask, "Why another policy statement on the environment? What purpose does it serve?" Let us therefore look at its unique features, then examine how the document may be of use to civil servants concerned with the environment, whether preparing to go to the negotiating table with international donors or trying to plan for the use of an area of marginal land in conjunction with other natural resources.

First and foremost, the "Environmental Perspective" is a statement formally adopted by the 158 member governments of the United Nations General Assembly in 1987, from the great powers to the smallest or poorest states. That so varied and vast a group could approve a document of this length on a highly controversial subject is in itself remarkable. Second, the title implies an unstated but all-pervasive word: *development*. It is a perspective for sustainable development. Implicit in its text is the assumption that environmental improvement requires examining the roots of environmental change even when these causes do not necessarily fall within the environmental purview.

The "Environmental Perspective" crystallizes the shared perceptions of the 158 governments. Fundamental among these are the beliefs that, in and of itself, poverty pollutes; that anticipatory and preventive policies are the most effective and economical in achieving environmentally sound development; and that the environmental impact of actions in one sector is often felt in others so that the internalization of environmental considerations in sectoral policies and programmes—and, equally, their coordination—is essential to sustainable development. In this last perception, echoed in the "Perspective's" examination of sectoral policies, the second major section of the document, lies much material environmental managers will find useful in helping traditional sector colleagues to shape their own programmes of action. For example, in agriculture and international trade, one has only to think of how environmentally sound irrigation can minimize the salinization of the soil and increase productivity and exports, including a broad spectrum of exports that may not be strictly agricultural but require some agricultural input.

From these shared perceptions flow goals in each development sector, as well as general recommended actions. It is worth noting, for instance, that in its discussion of industrial development, the "Perspective" recommends that the "Polluter Pays Principle" should not only be applied nationally, but extended to transboundary problems. This is a preventive action that both safeguards precious resources and minimizes costly damages and, in some cases, may well represent a source of revenues, internally and externally; the CFC and carbon dioxide emission charges currently in force in many of the industrialized countries are examples of applying this principle on a broad scale. Conversely, reward or incentive systems can be devised to bring about better implementation of sustainable development, which, to take but one case, reduce deforestation or, more generally, bring the actual and desired uses of land closer together.

Finally, the "Perspective" outlines five instruments of environmental action: assessment; planning; legislation and environmental law; awareness-building and training; and institutions. Again, all of these are cross-sectoral, notably the first two, reflecting the developmental tenor of the entire document. Similarly, awareness-building and training covers such occupational groups as engineers, builders, farm extension workers, and managers.

IMPLEMENTATION

In general, the developed countries have done far more to implement the recommendation of the "Perspective" than the developing nations—with the exception of Singapore, which has, in many instances, gone much further than many far richer states. Developing countries, of course, face enormous constraints in implementing these proposals. On the one hand, they face a far greater challenge to increase productivity in all economic sectors, among these, land use. While this drive may have environmental benefits, they tend to be incidental. On the other hand, many of their governments simply lack adequate data on which to base sustainable development planning and programmes. Despite these constraints, however, development goals can be stated in a variety of ways conducive to environmental protection. Likewise, social issues can and should be linked to the environment, although the farmer's need to earn a living involves short-term strategies, whereas environmental protection and enhancement generally require long-term action.

None the less, the two can be and have been productively reconciled. Wildlife policy in Zimbabwe provides a notable example. The communities living near wildlife preserves have been persuaded to carry out a number of conservation tasks because the government allots them a significant share of benefits of income earned from tourism and the export of wildlife products. Thus it has shown the feasibility of alleviating poverty, furnishing employment and raising income while protecting the environment. Moreover, this scheme has generated investment to manage the resources on which the livelihoods of these people depend. There are certainly other sets of circumstances in which comparable policies and programmes can be shaped. The factors for so doing are both minimal and immense: tremendous political commitment at the top and, on the part of civil servants like ourselves, imagination and the coordinated cross-sectoral elaboration of complex, highly detailed programmes.

DISCUSSION

Question: What is the relationship of the "Environmental Perspective" to the report of the World Commission on Environment and Development, also issued in 1987. They seem to make the same assumptions and draw many of the same conclusions.
Answer: There is a very close relationship indeed. The report of the World Commission on Environment and Development, published under the title *Our Common Future*, and the "Perspective" complement each other because they were written in tandem. In 1983, the General Assembly, well aware because of UNEP warnings, that the environment world-wide had deteriorated significantly since the Stockholm Conference of 1972, decided to improve the strategic policy approach of the United Nations. By its resolution 38/161, it established a special commission of eminent persons to look into the matter. However, having experienced disappointments with the implementation of the recommendations of comparable non-governmental bodies, such as the Brandt Commission, it wished to secure the commitment of the entire world community of nations. Consequently, it established an open-ended intergovernmental committee at the same time to keep in continuous contact with the World Commission and, based on its own political assessment of what was feasible and implementable, to prepare the *Environmental Perspective*. Both groups benefited significantly from their continuous exchanges.

Question: Is there any United Nations' directive that could bind developing countries to make available a minimal level of personnel and infrastructure on environment at the national level?
Answer: There isn't even a guideline on this subject—and it would be very difficult to make such a determination. In fact, very good results in environmental protection are brought about by people who do not necessarily work in environmental institutions.
Comment: In many countries, when other sectoral ministries know that there is a ministry or agency dealing with the environment, they become less concerned about this area. At the same time, those in the environmental ministry or agency cannot fulfil their mandate because of a scarcity of personnel, resources and infrastructure.
Comment: This is true not only of developing countries but of developed ones as

well. One way to circumvent some of the difficulties may be to keep a very low profile and attempt to work discreetly with colleagues in other ministries. The "Perspective" can be a tool in this regard, since it not only makes helpful recommendations for each sector but represents the commitment of every government that existed in 1987 to work in every sector towards enhancing environmental quality.

Comment: In offering aid, lending institutions and other donors often impose conflicting conditionalities.

Comment: If a developing country can demonstrate with figures and plans that it is making progress on the economic, social and environmental fronts in a coordinated way, it can build up significant strength for negotiation and establish a position of the mutuality of donor and recipient objectives. The fact that there is more donor money available for environmental activities than for others that may have greater national priority should not be regarded negatively or passively but as an opportunity for coordination within one's own government—for accelerated and sustainable development.

Comment: Such efforts are most successful where power, especially financial power, is delegated downwards, as far as possible to the grassroots level.

Chapter Four

National environmental management plans and programmes for sustainable development

MICHAEL ATCHIA, UTTAM DABHOLKAR, HALIFA DRAMMEH AND
MIKKO PYHALA
UNEP

INTRODUCTION

Until very recently in the field of the environment, governments seeking aid in this area were asked to prepare National Conservation Strategies. Since 1988, given the impact of the report of the World Commission on Environment and Development and the parallel "Environmental Perspective to the Year 2000 and Beyond", discussed in the previous chapter, as well as the Global Environmental Facility, the World Bank and other lending institutions, as well as individual donors, have requested the formulation of National Environmental Action Plans (NEAPs or, as they are otherwise known, National Environmental Management Plans) and have assisted some 50 countries in so doing. The funds provided for the drafting of such plans supplement the moneys allocated for project support and structural adjustment plans. Moreover, many donor governments have adopted specific budget lines for environmental planning and action. In addition, the Clearing-house Unit, the Environmental Law Unit, the Education and Training Unit and the Regional Office for Africa of UNEP can provide some further, albeit limited, assistance in the drafting of National Environmental Action Plans, particularly in support of workshops dealing with institution-building or strengthening, cleaner production programmes and those concerned with hazardous wastes. Contingent upon the development of plans in such fields, the United Nations Development Programme (UNDP) can assist governments with up to several hundred thousand dollars in the context of its country programming and implementation exercise.

This chapter is by no means a definitive exposition on National Environmental Action or Management Plans. Influenced largely by the experience of colleagues, consultants and resource persons who have worked intensively in the least-developed countries, notably in Africa, this essay's evolution and expansion will develop through the experience of forthcoming workshops, as well as further fieldwork in environmental management. By providing a checklist of questions and steps towards arriving at their answers, UNEP hopes that they will serve as a preliminary guide to this relatively new international emphasis on the integration of environment and development.

THE CHALLENGES OF SUSTAINABLE DEVELOPMENT

Unsustainable development has not generally been a conscious choice, but more the result of many factors including scientific uncertainty, administrative complexity and institutional inertia. Today, however, there is little uncertainty about the causes and consequences of unsustainable development. There is more than enough evidence and knowledge that countries can now make a conscious choice for a sustainable future.

However, choosing to move towards a sustainable pattern of development will require changes in many ways of thinking, deciding and doing, both within and outside government.

- *In thinking*, it means breaking away from old attitudes and approaches that treat environment and development as conflicting rather than as complementary and mutually supportive. Development that does not take environmental factors fully into account is simply not sustainable. Development that is not sustainable is simply not real development.
- *In deciding*, it means transcending the arbitrary boundaries between institutions that tend to be independent, fragmented and working on rather narrow mandates with closed decision processes. The interrelated and interdependent problems of environment and development require comprehensive approaches and more public participation, with new institutional arrangements and processes that integrate environmental and economic factors in all planning and decision-making.
- *In doing*, it means discarding or adapting practices, procedures and technologies that seriously degrade the environment or deplete natural resources, while also developing and practising new approaches that at the very least maintain and, preferably, improve the state of the environment and natural resource base on which human health and future economic development depend.

In most societies, these changes will require a transition period of years, perhaps even a decade, though the sooner they take place, the better for maintaining both economic development and political stability within and among countries. A pragmatic and planned process of change will be needed in which the exploitation of resources, the direction of investments, the thrust of technological development and the policies of major economic and sectoral agencies are reoriented and reconciled to enhance both the present and future capacity to meet human needs and hopes.

Those governments wishing to avoid current conflicts between environment and development will commit themselves to finding and implementing appropriate answers to the following fundamental questions:

1. "Where are we and where are we heading?"
Assess and report on the state of the environment and natural resources.

Review major economic and sectoral policies and assess their actual and likely impact on the environment and natural resource base and the consequences for future economic development. What capital losses, for example, have taken place in land degradation, freshwater pollution, forest damage or destruction, among so many other degradations?

Review and assess the adequacy of existing institutional and legal arrangements relevant to environmental protection and sustainable development.

2. "Where do we want to go?"
Set overall objectives and specific goals, guidelines and priorities for environmental protection and sustainable development.

3. "What should we do to get there?"
Prepare and publish a National Plan of Action and/or strategy for achieving sustainable development.

Undertake relevant education, training and public information programmes that reach the grassroots level on a basis of interchange between local people and the relevant authorities.

Strengthen the legal framework and formulate and enact effective norms and enforcement measures for the protection of human health and the environment.

Strengthen the decision-making process and institutional arrangements to ensure sustainable use of the environment and natural resources, including the link with relevant groups outside government. This last is essential if decisions and institutional changes are to be effective.

Modify (or adjust) the mandate and policies of the major economic and sectoral agencies to comply with the sustainable development goals, guidelines and priorities. These must be carried out with those who are knowledgeable about local, provincial and national environmental conditions.

Apply sustainable development guidelines to all new policies, programmes, technologies, products and projects—again, with participation at the grassroots level.

4. "What progress are we making?"

Monitor and report regularly on the state of the environment and natural resources and trends.

Report annually on the extent to which economic and sectoral policies and activities have contributed to the degradation, protection or improvement of the environment and natural resource base and the consequences for future economic development.

Prepare an overview report containing only basic information for most areas but highlighting one or two key environment and development issues in depth, with a plan for upgrading the information for the other areas on a priority basis in later reports.

As a further source of additional data and important information at the project level, countries can include the collection of environmental and natural resource data as an integral part of all projects carried out or supported by the government and require that environmental impact assessment be made a part of the feasibility studies for any projects likely to have an adverse effect on the environment or the sustainable use of a natural resource.

In moving from a preliminary to a more detailed and authoritative report on the state of the environment and natural resources and to assess later progress, many countries will need technical advice and assistance in applying appropriate monitoring and testing techniques and in obtaining equipment. The UNEP *Guidelines for the Preparation of National State of the Environment Reports* give a clear overview of what problems and parameters to assess.

However, information and advice on the technical means and options for doing so—and where and how to obtain essential equipment—will also be needed including:

- One or more guides on monitoring and testing techniques and technologies;
- A reference book on environmental monitoring and pollution control;
- A guide on how to improve national estimates of natural capital stocks of extensive resources (e.g. land, vegetation, surface water) through satellite imagery; and
- A special fund to help finance the purchase of essential environmental monitoring and testing equipment (in some cases this might lend itself to direct or coordinated bulk purchasing of equipment thereby lowering unit costs for developing countries). What role, for instance, is played by industry and large-scale agriculture aimed at foreign markets? Are environmentally harmful activities being taxed? Are environmentally beneficial activities being rewarded?

REVIEW AND ASSESS THE IMPACT OF MAJOR ECONOMIC AND SECTORAL POLICIES

The state of the environment and natural resource base is determined by the major economic and sectoral ministries. It is their policies and budgets and the decisions they take or avoid that determine whether the environment and natural resource base is degraded, maintained or improved. Although national development plans and agriculture, energy, industry, trade and other sectoral policies set economic goals and targets and the general guidelines and programmes for achieving them, they rarely involve or include an analysis of the actual or likely impact on the environment and natural resource base.

- What mechanisms exist for coordination and cooperation between national ministries and among different levels of government? How effective have they been?
- Are there any significant institutional or legal gaps?

These questions relate only to government. Since the transition to sustainable development will require the support and cooperation of many outside the government, the review should also address the following questions:

- What other groups and organizations have relevant roles and expertise (in the private sector, the scientific community and other non-governmental organizations)?
- What mechanisms exist for exchanges of information and views and for cooperation with them?

The review should also include at least some indication of external sources of possible expertise and assistance and address the following questions:

- What are the major bilateral and multilateral partners that could provide advice and assistance in helping make a transition to sustainable development?
- Which could best help in what areas?

Finally, the legal aspects of the review should also include relevant international law and address the following questions:

- What regional and global conventions and agreements have we ratified that contain provisions regarding environmental protection and natural resources management?
- Have we fulfilled our obligations under those conventions and agreements?
- What other relevant international conventions and agreements exist that we have not yet ratified but should consider?

Such reviews of legal and institutional arrangements have usually been presented in a narrative form which provides either too much or too little information for busy decision-makers to grasp quickly or to act on effectively. The text is often accompanied by one or more organizational charts showing the formal reporting structure rather than the functional relationships among different units and ministries. This approach needs to be supplanted or supplemented by one or more synoptic charts or matrices that provide a concise overview of key issues, functions and relevant laws and help pinpoint areas of overlap or duplication as well as significant gaps.

WHERE DO WE WANT TO GO?

1. The process for addressing this question in each country should ensure that the "we" eventually includes not only a representative cross-section of the best experts from different ministries and levels of government but also from the private sector, the scientific community and relevant NGOs.
   - The preparation and publication of an operational national environmental action plan, or programme and strategy for achieving sustainable development.

2. The plan, or programme or strategy should include specific measures to:
   - Strengthen the legal framework and put in place appropriate standards, regulations and effective enforcement measures for the protection of human health and the environment;
   - Strengthen the decision-making processes and institutional arrangements within the government to ensure sustainable use of the environment and natural resources and the links with relevant groups and organizations outside the government;
   - Adjust the mandates and policies of the major economic and sectoral agencies to comply with the sustainable development goals, guidelines and priorities; and
   - Apply the sustainable guidelines to new policies, programmes, technologies, product and projects.

3. In preparing and implementing their plans, programmes and strategies for achieving sustainable development countries can draw on relevant information, advice and assistance from many different bilateral agencies. However, there are so many international agencies with relevant programmes that have produced over 160 manuals, handbooks and checklists on environmental and resource management guidelines that countries may be intimidated by the sheer number and volume and be unsure as to how or where best to begin. There is a need to prepare concise and practical guides for policy makers and senior managers in the major economic sectors, each containing summary information and references on, for example:
   - Analytical tools and methods for assessing development options, making better decisions and evaluating their effects (e.g., cost-benefit analysis, impact assessment);
   - Major policy options (e.g., regulation, economic incentives and disincentives);
   - Sustainable development guidelines and procedures for applying them;

- Relevant international agencies and programmes.

4. The plan or programme or strategy can only be implemented successfully if it has the support and cooperation of an informed and consenting public. The government should therefore include as an integral part of the plan:
   - Special education and public information programmes on environmental protection and sustainable development (e.g. the National Environmental Action Plan of Seychelles was launched by its president on Independence Day, in the presence of representatives of the World Bank, UNDP and UNEP and given wide radio, television and press coverage);
   - Periodic consultation for an exchange of information and views with relevant groups and organizations representing the private sector, the scientific community and concerned non-governmental organizations and people (e.g. in the case of Rwanda, before its adoption by a National Workshop, the National Environmental Action Plan was brought to and discussed by villagers);
   - Governments, when presenting the annual budget, should also report on the previous year on the extent to which the economic and sectoral policies and activities have degraded, maintained or improved the environment and natural resource base and their likely impact on future economic development.

   A further important step, however, is to attempt to integrate the environmental and economic information for better economic and resource use planning and decision-making. For example, extensive research has already been carried out in Canada, France, Norway and the USA on techniques for changing the system of national accounts and reporting using both economic and physical units in a mixed accounting framework. Further work could usefully be done on how such new accounting frameworks linked to macro-economic tools such as national accounts could contribute to sustainable development in developing countries.

5. International agencies such as UNEP, UNDP and the World Bank could help in accelerating this process by providing expertise and assistance to:
   - Assess both the progress of on-going work on national accounting frameworks and the possible adaptation of the results for use in African countries;
   - Test such techniques on a pilot/demonstration basis in one or several African countries.

Based on these and other reports and studies, the government can ensure further progress towards sustainable development by taking further action to make appropriate adjustments in the national development plan and major economic and sectoral policies and make further changes to strengthen the relevant institutional and legal arrangements.

DISCUSSION

Question: Developing countries are often bewildered by the overlapping of concurrent exercises in planning requested by the donor community. Apart from the action plans called for by the international agencies, bilateral donors are asking for specific plans that frequently require totally different formats. Can UNEP suggest a way of reducing such difficulties?

Answer: In so far as possible, National Environmental Action Plans (or NEAPs)—which should include not only the input of the ministry or body responsible for the environment but all other ministries—should be the umbrella under which all other specific plans for the environment are formulated. Admittedly, there are significant and often divergent donor requests, but there have been a number of successes in fulfilling these, among them the NEAPs of Mauritius, Botswana, Zimbabwe and Seychelles assisted by UNDP, UNEP and the World Bank.

Question: World Bank structural adjustment loans involve conditionalities that affect national development plans, typically cutbacks affecting the implementation of environmental action plans that require the development of a strong civil service in this area. Doesn't this conditionality bring about a fundamental conflict?

Answer: At present, it is indeed a contradiction. However, it will doubtless be resolved as the process of formulating National Environmental Action Plans continues to grow. Their elaboration is increasingly becoming common practice in developed as well as developing countries. Even structural adjustment programmes now routinely require criteria of environmental management.

Comment: The World Bank insisted that our government formulate a National Environmental Action Plan, strongly suggesting that we create a centralized unit in the Ministry of Planning and Budgeting to examine sustainable development issues, including those concerning the environment. By contrast, a bilateral donor asked that every ministry include an environmental unit. Moreover, the World Bank insists that foreign experts implement the environmental programmes. We believe that donors should allow—indeed encourage—our local experts to take the primary responsibility for these programmes; they know the local people and the problems.

Comment: Perhaps the best solution for the long term is that developing countries strive to fund their environmental action plans internally rather than rely on donor contributions—and conditionalities—for their execution.

Comment: On the industrial taxation issue, many industries resist such charges for two reasons: first, they do not see their taxes being used for environmental rehabilitation programmes and, second, they bear the burden of developing the infrastructure needed for their activities.

Comment: One possible, even if partial, solution is to modify fiscal policies, such as eliminating subsidies to energy, water use and, in agro-industry, fertilizers. Apart from devising incentive systems to solicit environmentally friendly practices on the part of industry and large-scale agriculture, the savings from diminishing or abolishing such subsidies could be used for environmental enhancement.

Chapter Five

Environmental awareness, education and training

MICHAEL ATCHIA AND WIMALA PONNIAH
Environmental Education and Training Unit

AND

TORE BREVIK AND JOHN HARE
Information and Public Affairs Branch, UNEP

In 1972, the Stockholm Conference recommended that

> the Secretary-General, the organizations of the United Nations system, especially UNESCO, and the other organizations concerned, should . . . take the necessary steps to establish an international programme in environmental education, interdisciplinary in its approach, in school and out of school, encompassing all levels of education and directed towards the general public, in particular the ordinary citizen living in rural and urban areas, youth and adult alike, with a view to educating him as to the simple steps he might take, within his means, to manage and control his environment.

Further, Principle 19 of the Stockholm Declaration gave commensurate weight to the role of the mass media. Clearly, the goal was action.

Rising to this challenge in 1975, UNESCO and the then infant UNEP launched the International Environmental Education Programme (IEEP) with a seminar workshop of 20 experts in Belgrade. The charter they issued called for the development of "a citizenry that is aware of, and concerned about, the total environment and its associated problems and that has the *knowledge, attitudes, motivations, commitments* and *skills* to work individually and collectively towards solutions of current problems and prevention of new ones" (emphasis added). Again, the goal was action and the target everyone and the environment itself, as distilled by the educator, one of the chief media in this vast undertaking that encompassed education both as to how the environment affects human well-being, as well as how every detail of human behaviour affects the environment. Two further international meetings gave impetus to the development of the IEEP. The first, the Intergovernmental Conference on Environmental Education, held in Tbilisi, Georgia in 1977, issued a Declaration containing 41 recommendations for action at the national, regional and international levels that amounted to a global blueprint for the pro-

gramme. The second meeting was the UNESCO–UNEP International Congress on Environment Education and Training, held in Moscow a decade later, which produced the International Strategy for Action in the Field of Environmental Education and Training for the 1990s, the first part of which highlighted the increasing needs and changing priorities that had emerged since Tbilisi; the second, nine areas for further work, ranging from access to information through specialist training and international and regional cooperation.

As the first chapter of this volume has pointed out, environmental education for the 1990s is becoming distinctly more scientific and technical, with the strongest possible accent on application. As a recent issue of the quarterly IEEP environmental education newsletter, *Connect* (published in eight languages with a circulation of 20 000 individuals and institutions), stated:

> Two special groups of university students should be mentioned as target audiences for more intensive environmental education and training following that given to students in general. They are (1) student scientists, technologists and other future experts and professionals who will be dealing with environmental concerns (foresters, biologists, hydrologists, ecologists, agriculturalists and the like); and (2) those students of specific professions and social activities who will have an influence and impact on environmental management, both rural and urban, somewhat less directly (engineers, architects, urbanists, economists, labour leaders, industrialists, *et al.*).

Accordingly, since 1977, UNEP and UNESCO have sponsored a 10-month postgraduate course in environmental management at Dresden's University of Technology. For 1990 and 1991, a four-month intensive training course was set up at the Tufts University in the United States, offering a cross-sectoral approach to the management of natural resources. A number of universities in Third World countries, including Egypt, Kenya and Nigeria, are developing comparable undergraduate or postgraduate courses with UNEP help. This growing emphasis on scientific and technical education is also reflected in the fact that some 90 per cent of UNEP's current publications deal with highly technical subjects.

Further steps for the expansion and refinement of IEEP emerged from the 1992 United Nations Conference on Environment and Development in its Agenda 21 which placed emphasis, throughout its 40 Chapters, on capacity-building, i.e. awareness-building, education, training and institution-building.

IEEP is now into its ninth phase (1994–95), concentrating on four domains of action:

- Teacher education, including training seminars for primary and secondary instructors, as well as educational administrators and planners. Many proposals for this field were put forward at the 1990 World Conference on Education for All, held in Jomtien, Thailand, at which UNEP promoted the view that environmental literacy is an indispensable element of the campaign for universal literacy by the year 2000.
- Development of curricula and teaching materials: guides; handbooks and

resource books; case studies; prototype curricula for primary and secondary schools, as well as for teacher training, both pre-service and in-service; and teaching/learning modules, of which more than 30 have now been issued, covering such fields as the environmental problems of cities, of arid-lands, of oceans;

- Research and experimentation, including the adaptation of internationally developed materials to national needs; and pilot projects such as the incorporation of environmental education into university education;
- Environmental information, such as posters that encapsulate the major features of UNEP's concentration areas, as well as suggestions for activities to mitigate the salient problems of these fields.

Targeting is doubtless the most important element of all education and training and poses thorny questions. Given limited resources, who is trained in order to create the widest multiplier effect? Is he or she a key worker and/or role model? What is the relationship of the potential trainee to his or her co-workers? And who are these co-workers, simply subordinates or key workers in other fields with environmental implications? How will he or she disseminate what has been learned? Will this dissemination be limited to duplication and distribution of materials acquired during training or the development of a comparable course adapted to the needs of those with whom he or she works? Finally, is the trainee in a position to foster the development of a network for training? Whatever the answers to any of these questions, it should never be forgotten that the basic purpose of the training is a deeper understanding of the role of an enhanced environment in an enhanced quality of human life.

Targeting is also a prime concern of the public information specialist. In addition to publicizing UNEP's functions and achievements, the Programme's Information and Public Affairs Branch (IPA) alerts a wide variety of publics to pressing environmental issues and work being carried out world-wide to address them through its regular press features, its "green papers", environmental briefs on priority problems and more generally, through UNEP's popular magazine, *Our Planet*. The Branch strives towards several multiplier effects, one being the training of journalists, another the distribution of its many features to regional as well as international press agencies. UNEP has also played a seminal role in establishing the editorially independent news and information service, Earthscan, on which journalists the world over draw regularly. In the audiovisual realm, through the UNEP/Central TV-created Television Trust for the Environment, more than 60 international co-productions, reaching audiences in 80 countries, have been released.

Expanding the environmental participatory action of a variety of influential groups through its Outreach Network, IPA systematically maintains contact with special constituencies such as youth groups, women's groups, parliamentarians, religious organizations and industrialists. Each of its commemorations of World Environment Day, 5 June, is celebrated in a different country and concentrates on a different theme, encouraging public participation through art, photographic, musical and other competitions, as well as professional presentations geared to the year's particular focus. In 1987, to draw attention to the spectrum of environmental work being carried out at all socio-economic levels, UNEP launched a special award programme known as the Global 500, whose laureates range from high-profile personalities in politics and entertainment to ordinary citizens who have, in one way or another, improved the quality of their surroundings. And, looking to the future, its annual Global Youth Forum mobilizes today's young environmental idealists to lay the foundations for a network of tomorrow's environmental decision-makers.

Environmental awareness, education and training—all three are fundamental elements of capacity-building for sustainable development, keys to fostering an ethic of environmental stewardship. As many a traditional proverb maintains in one or another formulation, *Earth is less a legacy from our parents than a loan from our children.*

PART II

ECOSYSTEMS: PROCESSES, PROBLEMS AND STRATEGIES

Chapter Six

A brief overview of problems facing oceans, seas and coastal areas

PETER SCHRODER

Director, Oceans and Coastal Areas Programme Activity Centre, UNEP

Coastal areas are the most densely populated parts of the world: within 80 kilometres from the coastline live two-thirds of the world's population: 3.5 billion people in 1991 whose number will increase at least twofold in less than 30 years. Human settlements on coastal areas produce an enormous pollution problem. For example, 132 million people live in the coastal areas of the Mediterranean Sea—not a very high population level. However, to treat waste waters from their urban and industrial centres according to internationally acceptable standards would take approximately US$36 billion.

Coastal areas are heavily over-exploited with a consequent and immense alteration of local ecosystems. At least four elements are needed to ameliorate the situation; first, an institutional framework; second, there is the implementation issue: how to enforce environmental protection principles in the face of the general shortage of professional skills in this field. Third there is the need to raise public awareness and support to fulfil these needs. Fourth, there is the conceptual difficulty of tackling a broad range of issues that are interrelated. The only possible approach is an integrated coastal management planning.

This, of course, is not easy. A major problem is the intricacy and overlapping of jurisdictions, some stemming from traditional rights that conflict with a global approach: fishing rights and land rights.

The planning of integrated coastal management involves not only the impact of human activity on environment but also the occurrence of natural hazards such as cyclones which are unavoidable. Also, the development needs and well-being of the populations concerned require adequate consideration. The integrated assessment of these elements, however difficult, will be possible only if we have the necessary data and information. The general situation in this respect is disappointing, and there is a need for concerted efforts to improve the amount and quality of management data on which to base planning activity.

The impact assessment process is very complex because of the many interrelationships among activities in coastal areas, ranging from communalities through conflicts to synergetic effects. Yet it is necessary to define their impact in recognizable terms. Where does one start? Perhaps it is best to begin with the perceived priorities in those areas.

It is also important to define what a coastal area means, and this depends on the specific type of environmental degradation. If erosion occurs, problems of inland deforestation and watersheds are involved and inland activities have also to be considered. Fisheries may be less affected by such concerns but may be exposed to other types of problem such as off-shore oil exploration. Therefore each situation requires very specific assessments, not *ad hoc* treatment.

Let us list the most recurrent issues: fisheries, water supply, recreation, tourist facilities, port development, electricity production and associated with the latter the exploitation of energy resources such as oil, the development and supply of industries and agricultural development. What is needed to accommodate all these issues is that each situation be assessed to an adequate degree of specificity. For instance, in assessing the environmental impact of port development, it is important to define the scope and purposes of such activity: for containers, large oceanic vessels, cruise ships, oil tankers, and their respective requirements. Moreover, in considering these points, the factors of time frame and time frequency cannot be ignored.

To sum up, the organizational problem is a major issue. Lack of coordination among public agencies, departments or ministries needs to be addressed. Sometimes the establishment of a coordinating body is effective; sometimes assigning a leading role to an existing agency is preferable.

Other problems include lack of planning, lack of regulatory authority, inadequate knowledge and understanding of ecosystems; scientists are needed. Other issues are decisions that are based on purely economic grounds without environmental consideration; lack of clearly stated goals, lack of funds, short-term management approaches rather than long-term perspectives; complex, conflicting and confusing laws—when there are any at all. Lack of training, transfer of technology and limited public participation in decision-making are also obstacles.

To recapitulate: first, it is important to assess whether coastal development is an important priority for a country, *vis-à-vis* its level of development; sometimes it is not. Second, the dimensions of the problem must be determined. Finally, adequate data and properly trained staff are both indispensable to manage the resource sustainably.

DISCUSSION

Question: Is it really possible to segregate the management of coastal areas from the situation inland and related environmental issues?

Answer: It depends on the country's priorities—and, of course, the country's size. In some situations, there are almost total linkages, notably in island states. However, even if the island state is large, it may be possible. It is most important to choose key priorities and focus on them.

Question: It is very difficult to organize local protests or even raise awareness against excessive tourist facilities, especially when tourism is or can be their only source of income. How can one do so?

Answer: Effective action will vary from place to place, but we must convince local people and politicians that wrong decisions in the environmental context often produce bad economic results. Conversely, environmentally sound arrangements are usually beneficial economically in the long term, so that both tourism and the welfare of local inhabitants can co-exist.

Question: How can one bridge the gap between environmental experts and decision-makers and planners?

Answer: Again, the solution will vary from one country to another. But you might identify a powerful, relevant, high government official, such as the Minister of Finance or Planning and ask him or her to designate a focal point. Strong political support is always needed for environmental initiatives.

Chapter Seven

Elements of marine science and coastal ecosystems

GEORGE E. KITAKA
Programme Specialist for Marine Science, UNESCO/ROSTA

INTRODUCTION

In view of the vast expanse of the world oceans, which cover almost three-quarters of the earth's surface, and the seemingly inexhaustible resources they contain, the importance of scientific investigations of the oceans cannot be over-emphasized. In many cases, these investigations concern practical interests and uses, such as navigation, coast protection, the production of raw materials and food, or the conservation of the coast for recreation in the face of potential forms of degradation or marine pollution. In other cases, however, the investigations are purely scientific in character and may be directed towards a better understanding of the characteristics and functioning of the entire ocean ecosystem. In this case, they are referred to as 'oceanography'. When they concern the various component ecosystems in different parts of the sea, they are collectively termed 'marine science'. From this loose demarcation, it may be concluded that whereas oceanography would qualify as a marine science, the converse does not necessarily follow. What is common to both terms is that they are collective and refer to a multidisciplinary science.

It is also important to note that whereas investigations of the first category could be limited to the scientific discipline or disciplines most directly applicable to the need in question, those of the second category must include the study of the whole sequence of geological, chemical, physical and biological events that operate together in the ecosystem under investigation and that are interdependent.

THE COASTAL ZONE AND ITS ECOSYSTEMS

One of the areas of the sea that is of particular significance to civilization, and which is also of special relevance to the theme of this volume is the coastal zone.

Coastal areas are the site of complex natural systems in which intense interactions occur between land, sea and atmosphere. They comprise a variety of lucrative ecosystems that differ in nature, magnitude and importance from one coastline to another. They include creeks, estuaries and deltas, coastal lagoons, coral reefs, mangroves, and seagrasses, to name the most common. They differ in various ways from one coastline to another. For instance, whereas coral reefs are a characteristic feature of the East African coast, mangroves and coastal lagoons dominate the West African coast. Estuaries and deltas are common to both.

Generally, marine coastal ecosystems are highly productive, as they contain a wide variety and high abundance of living and non-living resources. They provide nursery grounds and shelter for a number of marine species, and play a major role in the world's fisheries. More than four-fifths of fishery catches come from coastal regions. Some systems, such as mangroves and coral reefs, also offer physical protection of the coastline from erosion.

The living and non-living resources of these environments yield a wide variety of products, ranging from fuel to construction materials and from food to medicines. A number of mineral resources are being exploited in the coastal zone. For instance, a substantial portion of the world's oil reserves is to be found in the coastal and continental-shelf areas. Yet coastal ecosystems are extremely vulnerable, particularly to increasing human activities in the coastal areas.

To elaborate further on the importance of two of these ecosystems, let us take mangroves and coral reefs: mangrove forests are considered one of the most productive and biologically diverse ecosystems on Earth, supplying important habitats for several species of fish, invertebrates and plants. Their root system provides sanctuary for sponges, crested worms, crustaceans and molluscs, as well as macroalgae. Intertidal zones create habitats for a variety of crabs and small animals, while hundreds of species of birds nest in mangrove canopies. Mangrove estuaries shelter marine mammals such as dugongs and otters, as well as certain reptiles such as the Indo–Pacific crocodile. It has been estimated that one hectare of mangrove forest, if properly managed, could produce an annual yield of 100 kg of fish, 25 kg of shrimp, 15 kg of crabmeat, 200 kg of molluscs and 40 kg of sea cucumber. In addition, the same area could supply an indirect harvest of up to 400 kg of fish and 75 kg of shrimp that mature elsewhere.

Coral reefs are another important ecosystem that rivals tropical rainforests in species richness and diversity. They, too, are highly productive. In regions in which they are found, nearly one-third of all fish species live on these reefs, while others depend on reefs and seagrass beds for various stages of their life cycles. An important factor to bear in mind is that reef-building corals depend on sunlight for their efficiency. They cannot reproduce in turbid water. Unless the water is exceptionally clear, most coral growth stops at a depth of 20 metres.

HUMAN IMPACT

Since early times, human settlements and, indeed, civilization have tended to develop in coastal areas. For centuries, these human societies had constituted an integrated and balanced component of the coastal environment. Social practices were often aimed at the protection of the settlement's natural surroundings.

Now, however, more than two-thirds of the world's population, and the majority of its largest cities, are located within 80 km of coasts, estuaries and deltas. The emergence of high-energy-based technologies and the rapidly developing industrial, transport, residential and recreational complexes have seriously threatened the equilibrium of the coastal environment and its ecosystems. For instance, in several parts of the world, mangroves are in retreat throughout their range. Clear-cutting for timber, fuelwood, and outright destruction for the creation of brackish fish and shellfish ponds, as well as for the expansion of urban areas and agricultural lands has claimed millions of hectares globally. Coral mining and blast fishing have proved particularly destructive, since most coral species grow very slowly, while exessive siltation from rivers has choked to death large areas of coral reef in a relatively short time.

Furthermore, being at low altitudes, the coastal zone and its ecosystems are influenced by what happens on land. Much human refuse ends up sooner or later in coastal waters, washed down by rivers or dumped directly. Rivers also bring into the sea billions of tonnes of sediments in which there may well be unknown quantities of poisonous residues from agricultural chemicals and heavy metals from industrial discharge.

COASTAL ECOSYSTEMS MANAGEMENT

To quote from Don Hinrichsen's *Our Common Seas*:

> There are no easy solutions to the human and resource crises afflicting the world's coastal areas. Land and sea must both be managed in a way that permits economic development, yet sustains the resource base. This involves the balancing of a multitude of human uses with one another, as well as managing resources in such a way that future needs are not sacrificed for the expediency of the moment.

I would only add that such a strategy depends on management decisions that must be based on a proper understanding of the functioning and mutual interaction of the natural systems concerned. It is important to know, through research, at least the main aspects of such systems such as their ecological characteristics and functioning; their interactions and exchange of energy and material; and their relations with sea and land, for which properly planned interdisciplinary marine science programmes are essential.

DISCUSSION

Question: What is UNESCO doing to conserve mangrove and coral reef habitats?
Answer: UNESCO's programme is directed to the study of marine coastal systems. Under this programme, the Organization promotes scientific studies of all important ecosystems. The choice of the ecosystem to be studied in selected countries is made by their respective governments. As to Africa, UNESCO has selected certain countries in which to study such ecosystems as mangroves, coastal lagoons and

coral reefs. The Organization has organized many workshops and publishes articles and newsletters to promote public awareness.

Question: What steps have been taken to combat marine pollution caused by oil tankers?

Answer: Marine pollution is addressed mainly by UNEP, the International Maritime Organization (IMO) and the Intergovernmental Oceanographic Commission (IOC) of UNESCO. Much has been done in terms of sensitization, but there is little more that these organizations can undertake because it is difficult to police oil tankers and difficult to contain the results of accidents, which cause the more serious cases of pollution.

Question: What are the major causes of the destruction of mangroves?

Answer: Mangrove forests retreat largely because they are physically cut down by human beings. Consequently, people must be educated about the beneficial contribution of mangroves to the environment. However, other causes are also involved in this destruction, notably industrial and marine pollution.

Question: How significant to the pollution of coastal waters is the pollution of water bodies from the mainland?

Answer: It depends on the course these pollutants take. Some are not degradable, like DDT. Others break down during transportation. Depending on the course of the river, some pollutants can be removed well before they reach the coast.

Question: Has any work been done on the presence of mineral deposits, notably oil, along the East African Coast?

Answer: It is difficult to know. This type of activity falls under the jurisdiction of governments, which contract exploitation companies. The results of their research are rarely publicized.

Chapter Eight

Freshwater resources: a brief introduction

JORGE ILLUECA
Former Coordinator of Environmental Management, UNEP

Freshwater resources are the most precious of the earth's physical resources and the basic ingredient for supporting terrestrial life systems. It is estimated that 1386 million cubic kilometres of water are available on our planet. Of this total, only 2.6 per cent is fresh, the remainder being sea water or brackish. Therefore the volume of freshwater is estimated at only 37 million cubic kilometres, sufficient to cover the Earth's land surface to a depth of 250 metres.

However, very little of this amount is easily available to use in liquid form. Icecaps and glaciers contain 76.5 per cent of freshwater and another 22.9 per cent is found as groundwater. This means that less than 1 per cent of freshwater (six-tenths of a percentage point, to be exact) is available in the atmosphere, streams and lakes.

After subtracting runoff from continents into the sea, the World Resources Institute in 1987 estimated that only 9000 cubic kilometres of freshwater are readily available for human exploitation world-wide. Theoretically, this is enough water to support a total world population of 20 billion people. However, human population and the resource itself are distributed unevenly. Consequently, in some areas tremendous pressures are placed on the resource, while in others only a very small percentage will be exploited. Please bear in mind that these figures are estimates and will differ from one source to another, since measuring freshwater resources on a global scale is no easy matter.

Regions of the world that are currently facing shortages of water include Africa, the Middle East and parts of western North America. By the year 2000, many countries in the world will be besieged by a scarcity of water resulting from the increasing demand for water for agriculture, industry and domestic consumption. It has been estimated that the total use of water has increased tenfold between 1900 and the year 2000 or from approximately 600 cubic kilometres per year in 1900 to the nearly 5500 cubic kilometres projected for the year 2000.

By the year 2000, the exploitation of water will be divided in the following manner: 69 per cent for agriculture, 23 per cent for industry and 8 per cent for

Table 1 Estimates of Average Annual Streamflow of Freshwater by Continents

	Average streamflow (Km^3/year)		
Continent	Baumgartner & Reichel	Shiklomanov	L'vovich
Europe	2 800	3 210	3 110
Asia	12 200	14 410	13 190
Africa	3 400	4 570	4 225
N. America	5 900	8 200	5 960
S. America	11 100	11 760	10 380
Australasia & Oceania	2 400	2 300	1 965
Total	39 700	44 540	38 830

domestic uses. The principal use of water is for agriculture. World-wide, an estimated 235 million hectares of farmland are under irrigation.

Pressures on the resource have been felt most heavily on aquatic ecosystems. Rivers have been dammed and their waters diverted; many lakes are diminishing in size as the waters of feeder streams are diverted for agriculture and industry. The over-exploitation of groundwater, particularly in arid and semi-arid areas, is resulting in the depletion of this resource that is not easily renewable. In coastal areas, the exploitation of groundwater has often resulted in the intrusion of saline waters into aquifers.

While increased demands on the resource have impacted heavily on natural aquatic ecosystems, they have also led to the creation of artificial aquatic ecosystems-reservoirs. By 1986, 36 327 large dams (in excess of 15 metres in height) and with a combined capacity of 5500 cubic kilometres had been built, with another 1026 under construction. Much of this development is taking place in the Asian and Latin American regions.

Up until now, I have focused primarily on the issue of water supply and extraction. However, an equally serious problem is the deterioration of water quality that has accelerated during the past three decades. Both surface and ground waters, the latter more recently, have suffered serious deterioration of quality in many countries.

The discharge of untreated or inadequately treated waste water into rivers, lakes and other aquatic ecosystems is the principal source of pollution. Industrial waste waters in particular have wreaked havoc on aquatic biota in many areas. Runoff of agrochemicals from farmlands has also contributed to the eutrophication of lakes and reservoirs and has also killed aquatic biota. Untreated sewage water not only poses a threat to human health but is also contributing to eutrophication and the destruction of aquatic ecosystems.

A more recent problem that has received much attention during the last two decades is the acidification of lakes, when results from acid rains produced by atmospheric pollution, particularly in industrialized regions such as North America and Europe. In a 1986 study, it was estimated that 4600 out of approximately 83 000 lakes in Sweden had suffered severe acidification (pH values equal to or less than 4.9). Four thousand lakes were devoid of fish and another 17 000 had suffered reduced populations of acid-sensitive species. Most of these were small bodies of less than 1 sq. km.

Correcting the deterioration of water quality—which in many countries aggravates the problem of water scarcity—requires costly monitoring, treatment, control and regulation measures that often are beyond the reach of developing countries. Solutions are complex, requiring additional funding, transfer of technology, personnel, training, and re-equipping and reorientation of the industries that are the sources of pollution.

Reaching solutions for problems of water supply and quality is usually more difficult if the resource is shared by more than one country. Without taking into account the recent and ongoing redrawing of borders in Central and Eastern Europe, there are 214 international river and lake basins in the world. Of these, 155 are shared by two countries; 36 by three countries; and 23 by four to 12 countries. In approximately 50 countries, 75 per cent or more of the national territory falls within an international water basin. *Close to 40 per cent of the world's population lives in international watersheds.*

In order to address adequately the issues of water scarcity and deterioration of water quality in these basins, a collaborative effort is needed among the countries concerned. Such efforts are frequently obstructed by ongoing political differences such as border disputes, poor relations, competition for the resource and differences in social and economic goals. As demands grow on freshwater resources, countries will tend to perceive the management of this resource as a security matter.

For this reason, UNEP has given a high priority to the preparation and implementation of integrated management action plans for international watersheds which we feel will greatly increase in importance during the next 30–50 years. It is only through a collaborative effort among countries that the serious problems of water scarcity and deterioration of water quality can be effectively tackled. The unfortunate alternative in some cases could be the use of armed force to secure control over this precious resource.

DISCUSSION

Question: In view of the increasing requests by companies to buy fresh water from governments, what can be said about freshwater marketing?

Answer: The difficulties of obtaining freshwater are expected to increase significantly during the next 20–30 years and to become a global problem. Some specialists predict that the world will confront a substantial freshwater shortage by the year 2030; this will entail severe development limitations. Consequently, one can reasonably envisage an increase in freshwater shipping via tanker in the near future. Indeed, importing freshwater may become very expensive for some countries. It should be noted that a great deal of work has been done in the desalinization of sea water. However, as the costs are prohibitive for most countries, this approach may remain only a temporary solution to the shortage of freshwater. In fact, it is not unrealistic to assume that by the 21st century, freshwater may be as expensive as oil, the more so as huge freshwater sources, such as glaciers and ice caps, are not readily accessible.

Question: What is UNEP doing about wetlands, which are so often ruined by exploration for oil?

Answer: In this area, UNEP has taken an active role, but any UNEP field activity depends on requests from governments, which often perceive the control of freshwater as a sovereignty and, increasingly, a security issue. We have indeed been dealing with wetland ecosystems—largely, however, playing a support role *vis-à-vis* other agencies. In UNEP, we deal with water management at the integrated level. This calls for close coordination with all international bodies in this field.

Question: Has UNEP made an environmental impact statement about the ongoing extraction of underground water in Libya?

Answer: National assessments cannot be undertaken without the specific request of member states. In this particular case, UNEP has not been requested to provide its views on the project. But we are very much concerned because large-scale exploitation of groundwater may set off geological chain reactions that require careful study. In addition, the question of how long the water will last is fundamental. When we prepare an environmental impact assessment, we consider not only the protection of the environment but also the project's impact on the sustainable development of the country. Therefore, we address not only environmental issues but the social and economic effects of the exploitation of natural resources. The entire picture—and nothing less—is what we mean when we speak of sustainable development.

Question: Freshwater pollution, whether industrial or agricultural, often leads to the proliferation of certain species, even to the point of infestation. How should this problem be tackled?

Answer: Solutions will vary with the particular problem, but the best strategy is to look for biological control of the species in question. It should be added that a major direction in environmental management during the last decade has been strengthening assessment through the Global Environmental Monitoring System (GEMS), particularly in the areas of water quality supply and *species*. However, this kind of activity calls for far greater efforts by the entire United Nations' system.

Chapter Nine

Efforts to combat land degradation

TIMO MAUKONEN
Senior Programme Officer, Desertification Control Programme Activity Centre,
UNEP

INTRODUCTION

The ever more pressing problem of desertification and land degradation was understood long before the 1972 Stockholm Conference on the Human Environment and the establishment of UNEP.

A concerted effort to fight this problem at global level resulted in the 1977 United Nations Conference on Desertification (UNCOD), which took place in Nairobi. This conference produced a comprehensive global Plan of Action to Combat Desertification (PACD), which included 28 recommendations covering all fields related to the desertification problem. UNCOD also made proposals for institutional arrangements and financial mechanisms for combating the desertification scourge: mechanisms such as the United Nations Consultative Committee on Desertification (DESCON), the Inter-Agency Working Group on Desertification (IAWGD), and the UNEP/UNDP Joint Venture with the United Nations Sudano–Sahelian Office (UNSO).

Maybe the PACD was premature in its comprehensive and exhaustive character, too complex, for the world to understand and accept. With the wisdom of hindsight, many experts have found that it lacked the focus necessary to bring about the implementation of the programmes it recommended. Unfortunately, too, the mechanisms created by UNCOD failed to provide a significant spur to desertification control. None the less, local and individual success stories did emerge from certain approaches, among them the following:

- several training programmes;
- UNSO work;
- a few national plans of action to combat desertification;
- a few transnational projects;
- various bilaterally assisted and implemented programmes; and
- some monitoring exercises.

Perhaps another reason for the failure to realize the goals of PACD lay in its emphasis on the very word "desertification" which, to most people, connotes an image of the Sahara expanding. The clarification of terminology is therefore vital.

THE PHENOMENON OF DESERTIFICATION

Desertification is a widespread process of land degradation throughout the drylands. It differs significantly from the phenomenon of observed cyclic oscillations of vegetation productivity at desert fringes ("desert expansion or contraction") revealed by satellite data and related to climate fluctuations. These oscillations may affect the desert fringes within a band of several hundred kilometres—for example, seasonal expansion on the southern fringes of the Sahara. Consequently, such terms as "desert expansion", "desert creep", "desert advance" are not applicable to the phenomenon of desertification and are grossly misleading.

Of course, at the desert fringes it is very difficult to distinguish physically between desertification, a long-term degradation process, and a periodic reduction of vegetation productivity related to the temporary climatic change, such as a reduction of annual precipitation, which may be reversed in the few years that follow. None the less, all three phenomena: desertification/land degradation, vegetation productivity oscillations at desert fringes ("desert expansion or contraction") related to climate fluctuations and the recurrent droughts in drylands are closely interrelated. Perhaps we should think instead about the protection and development of drylands.

A recent UNEP assessment of the 1990–1991 status of global desertification shows the extent and degree of land degradation in drylands of the world by continents and by major land-use systems such as irrigated croplands, rainfed croplands and rangelands. For purposes of this assessment, UNEP defined desertification as *land degradation in arid, semi-arid and dry sub-humid areas resulting mainly from adverse human impact*. In this concept, land includes soil and local water resources, land surface and vegetation or crops. Degradation means the reduction of resource potential by:

- one or a combination of erosion and sedimentation processes acting on the land (water erosion, wind erosion and sedimentation);
- long-term reduction in the diversity of natural vegetation and crop yields; and
- soil salinization or sodication.

Currently, desertification directly affects about 3.6 billion hectares—70 per cent of all the drylands of the world, or nearly one-quarter of the total land area of the world. It also affects one-sixth of the world's population. These figures include such areas as the Sahara in which desertification seems to have reached its limits.

Desertification in the drylands manifests itself through:

- over-exploitation and degradation of 3333 million hectares—about 73 per cent of the total area of rangelands;

- decline in fertility and soil structure leading gradually to soil loss in 216 million hectares of rainfed croplands—nearly 47 per cent of such croplands in the drylands;
- degradation of 43 million hectares of irrigated croplands—nearly 30 per cent of such croplands in the drylands.

Currently, global financial loss (income foregone) due to desertification is estimated at US \$42 billion annually. This figure represents only direct on-site loss. Indirect off-site and social costs of desertification damage may be two to three or even 10 times as high.

The cyclic oscillations of vegetation productivity on desert fringes present the main obstacle for defining and plotting exact desert boundaries both at global and continental scales. The geographers are still arguing about the definition of desert because the traditional use of the term, being applied to both the nearly lifeless Sahara and the relatively green Arizona or Kara-Kum and Gobi, does not facilitate any precise scientific definition in measurable parameters.

DESERTIFICATION CONTROL

Desertification/land degradation is largely a human-induced problem in the drylands. Human beings can solve it if they properly understand and tackle it. It is not like going against the forces of nature. If anything, it is the opposite. Combating desertification is the restoration of natural ecological balance in the drylands. It is no more complicated that—technically.

However, desertification comprises a host of social and political problems. First, all the resources at our disposal—even the renewable resources—are limited and the number of mouths to be fed world-wide is increasing. Second, the distribution of wealth and resources is tremendously uneven, not only between continents and nations but—even more so—among the citizens of a given country. Those living in the less-endowed drylands do not generally receive the political attention they merit and therefore become more a source of exploitation than a target of investment. Nor have they been offered alternative sustainable livelihoods. Third, the general land-use conflicts arising from increasing population pressure, as well as questions of land tenure, need to be solved before many technical options can be applied and appraised for feasibility. Here the political will for compromise comes to the test.

In addition, internationally, desertification control has not been a primary interest of the donor community—partly because of the misconception of the phenomenon of desertification discussed above and, partly and more importantly, because the programmes and projects prepared for desertification control at the technical level of national ministries (with or without assistance from outside) have very seldom reached the priority lists of governments, that is to say, the ministry of finance or foreign affairs when these ministries enter into negotiations with donors. Also, only in very few cases are national plans of action to combat desertification incorporated into national economic and social development plans.

This development problem is also manifested in many of the ill-conceived projects prepared by outsiders unfamiliar with the target areas who do not seek the opinions and inputs of the target group and incorporate these into the project design. More often than not grassroots people do not participate at all in the planning process and find themselves somewhat abruptly asked to carry out work for a project that has little to do with their perceived needs. Unfortunately, desertification control has tended to follow this pattern, though lately the inclusion of villagers and non-governmental organizations in the project design and implementation has become far more frequent.

THE FUTURE

The United Nations Conference on Environment and Development devoted much attention to desertification, and its Agenda 21 comprises five main programme areas for combating desertification and drought:

1. Strengthening of the knowledge base and monitoring systems.

2. Combating land degradation and intensifying afforestation/reforestation.

3. Designing integrated programmes for poverty eradication and alternative livelihood systems.

4. Developing and integrating anti-desertification programmes into National Environmental Action Plans (NEAPs) and National Development Plans.

5. Developing drought preparedness.

However, agendas and political manifests and statements, necessary as they may be to catch the world's attention, *are not enough*. Nothing really meaningful will happen unless the people in the affected areas are involved in the development of their own resources sustainably. This means that their rights and their needs must be recognized, respected and cared for. They may well be too powerless to fight for their own cause; and that may well be the most important task of the environmental manager.

DISCUSSION

Question: Is data on desertification available on a national rather than continental basis?
Answer: In a number of cases, countries were unable to provide the data requested; the global assessment of desertification consequently suffered. But Agenda 21 of the United Nations Conference on Environment and Development emphasizes the necessity of gathering such information.

Question: Would a reduction of the pressure on biomass as fuel reduce the desertification process?

Answer: Undoubtedly. New forms of energy could help save the grasslands, but nuclear and solar energy are only a long-term prospect. For the present, greater use should be made of animal traction rather than diesel oil to pump water and to perform other functions.

Question: Why has the question of desertification been eclipsed by other environmental issues?

Answer: It is a matter of political will. But it cannot be completely overlooked because it is so closely related to climate change, biological diversity, forests and other currently high-profile issues. Nevertheless, it cannot be forgotten that desertification is in large measure a political and social problem. Politicians tend to respond to the demands of urban populations which demonstrate. The rural dryland peoples are marginalized, powerless; they lack the will to make their voices heard; they don't know how. Environmental managers can help voice their problems.

Chapter Ten

National soils policies for the protection and rational utilization of soil resources

ALI AYOUB
Soils Unit, UNEP

INTRODUCTION

Let us try to think of crop cultivation as a kind of mining. When any crop is grown, essential elements are taken from the soil. Generally, only three are replaced: nitrogen, phosphorus and potassium. Unless a far more balanced mixture of nutrients is supplied, the soil becomes depleted. This is happening today world-wide at an alarming rate, largely because of this century's demographic explosion, as well as a revolution in expectations created by the images of a better life seen daily by many of the world's poor, whether with their own eyes in their own surroundings or through the mass media.

These two forces, working in synergy, are increasing the demand for agricultural products. Given these pressures, as well as others, soils are being lost or degraded in a number of ways, notably the following:

- physically lost through accelerated erosion;
- degraded by such processes as the accumulation of salts (as in salinization and alkalization); progressive leaching and acidification; loss of organic matter and of soil structure; and waterlogging or the unwise addition of chemicals;
- in addition, millions of hectares of good farmland are being lost each year to non-farm use for urban and industrial development and other purposes.

Because of all these factors, the formulation and implementation of national soils policies has become a matter of the greatest importance and urgency.

PRINCIPLES OF A SOILS POLICY

A national soils policy should allow and stimulate maximum utilization of the soil on a sustained basis without lowering its productivity and without causing direct or indirect damage to the environment. The implementation of the policy should be taken into account when any aspect of national development is being considered. The objectives and lines of action associated with the basic principles of a national soils policy will, to some extent, vary from country to country according to the natural environment, and to social and economic factors. In broad terms, however, national soils policies should set out to:

- Assess available land resources in terms of quantity, quality and liability to degradation;
- Improve the productivity of soils by applying scientific knowledge and better management techniques and by developing and promoting more productive agricultural systems that assure the use of the soil on a sustained basis;
- Enlarge the area and improve the quality of available agricultural land wherever feasible through irrigation, flood control and reclamation;
- Slow down the loss of productive agricultural and forest land to industry and other uses;
- Monitor changes in soil quality and quantity and monitor the way land is used;
- Bring to the attention of all concerned the dangers and adverse consequences of soil degradation and the need for conservation and appropriate legislation.

National institutions should also be created or improved to realize these objectives.

SOCIAL AND ECONOMIC ASPECTS OF ARRESTING SOIL EROSION AND DEGRADATION

Virgin soils are often thought of as having accumulated over a long period of time reserves of nutrients that are then exploited or mined by the cultivator. Although, in some cases, progressive farmers have over the years built up the chemical and physical properties of their soils by careful husbandry and the use of fertilizers, the usual picture is one of exploitative cultivation that progressively uses up the nutrient reserves of the soil, adversely affects its physical properties, such as structure and porosity, and leads to an overall decline in productivity. The rate of this decline depends both on the intrinsic quality of the individual soil (since some are inherently much more "fragile" than others) and on the way the soil is used.

Sooner or later, most cultivators are faced with the choice of either putting back into the soil at least some of the nutrients removed by their crops, as well as of adopting appropriate conservation measures to prevent further soil erosion—or of seeing the productivity of their land decline still further. The cultivator's decision as to when he is going to do this—and how—is related to his estimate of the costs involved in relation to the value of the yields he is losing and the declining capital value of the land. Short-term improvement is economically essential when the annual decline in capital value of the land exceeds the net annual income obtained by exploitation of the soil, since from that point onwards the farmer, in real terms, is losing income. However, it is not easy for the small farmer to see when that point is reached because of the difficulty in estimating the monetary value of the decline in the capital value of his land which is often hidden by an inflationary rise in land values.

Even if the small farmer is conservation-minded and sees the need for investment, it is doubtful that he has the knowledge, initiative and energy necessary—or the capital—to plan and execute satisfactorily the role of the state in soil conservation and improvement. The continued short-term exploitation of land may be profitable to the farmer at the time, but it is not profitable to the farmer all the time and is not profitable from the point of view of the community as a whole.

The extent and need for state intervention in soil conservation has many aspects. Technically, the state should be in a position to give advice and help in planning conservation measures. Financially, loans subsidies or tax relief may be necessary. Economically and socially, one has to consider whether practices that are too costly for the individual may nevertheless be relatively inexpensive for the society as a whole—even essential for its future welfare. The unique position of the state with respect to finding capital, covering risks and finding technical solutions in soil conservation must be recognized.

State action in the field of soil conservation, if it is to be based consistently on sound principles, should take place within the framework of an agreed soils policy. Economic, social and legal conditions should be established to stimulate and encourage the use of the soil on a sustainable basis.

In a fundamental sense, a farmer who produces a crop and sells it is exporting nutrients from his land, even if he is not losing soil through accelerated erosion at the same time. In the same way, an agricultural-exporting country is exporting plant nutrients just as an oil exporter is exporting reserves. In the long run, the ability of the farmer, or the exporting country, to replace those nutrients in the form of fertilizers, as well as to pay for the conservation measures necessary to reduce soil erosion to acceptable proportions, depends on the selling price obtained for this produce. In terms of agricultural exports from developing countries to the more developed, or in recent terminology, from south to north, much depends on the prices of the products *vis-à-vis* those of manufactured imports. In this regard, fair prices for agricultural exports must be seen as essential if the soils of the agricultural-exporting regions of the world are to be kept productive. To pay too little is disastrous in the long run, a fact recognized at the United Nations Conference on the Human Environment (Stockholm, 1972).

However, in some cases, the prices paid to a country for its agricultural exports never reach the farmers concerned because the state, often through a state-run produce marketing or exporting board, passes on to the farmer only a fraction of the price received. Such policies, however attractive they may seem in the short term, are nevertheless disastrous in the long run, as they lead to deterioration in the productive capacity of the land.

WATER MANAGEMENT AND CONSERVATION

Water is an integral part of the soil/land productivity base. The misuse of water can lead to soil degradation or erosion. Good water management is necessary for increasing crop yield. New irrigation schemes require a high and long-term investment and should be based on a good and scientific knowledge of the soil–plant–water relationship.

TECHNICAL ELEMENTS OF NATIONAL SOILS POLICIES

Types of technical elements

Technical elements of a national soils policy will necessarily vary to some extent according to local conditions and the degree to which national services supporting agriculture have been developed. However, in many cases they include the following three groups of aspects:

- Those relating to soil inventory, soil assessment and land-use recording;
- Those relating to protecting the soil from erosion and degradation, and from alienation for non-rural use;
- Those relating to soil management (including appropriate farming systems), soil improvement and soil reclamation.

These groups comprise elements that are not static, but subject to change. Since they change with time, there is a need to monitor the changes taking place. Monitoring should reveal current trends and problems, allow predictions to be made and appropriate actions to be taken in response.

The three groups of technical aspects might usefully be further sub-divided as follows:

1. Soil mapping, classification and assessment, and land-use recording
   These aspects include:
   - Agreement on and adoption of an internationally agreed system of land evaluation and soil taxonomic classification, in addition to local systems of classification (if already established);
   - The adoption and elaboration of systems of land evaluation and land capability classification, particularly those which draw attention to conservation needs and stimulate utilization of land according to its capability;
   - The recording and monitoring of land use and farming systems.

2. Soil protection from encroachment and degradation
   Technical aspects of soil protection include:
   - Research on soil erosion and on measures for its control, including research related to the protective aspects of improved farming systems;
   - The monitoring of soil degradation;

- The implementation of technical measures designed to control degradation;
- The protection of agricultural land from alienation for non-rural purposes.

3. Soil improvement

Technical aspects relating to soil improvement include:

- The improvement of soil physical and chemical properties by the adoption of appropriate farming systems including tillage methods, rotations; and the use of residues, fertilizers and amendments;
- The introduction of irrigation and drainage;
- The rehabilitation of soil and land that has been disturbed or degraded;
- The reclamation of saline, flooded and other land not at present productive, based upon both social and/or economic considerations.

INSTITUTIONAL ELEMENTS OF NATIONAL SOILS POLICIES

Functions and types of institutions

The institutions concerned with soils and soils policy that now exist in different countries, or that would be desirable, vary very considerably from country to country. Broadly speaking, the human institutions needed to implement the elements of a national soils policy can generally be thought of as fulfilling the following main functions:

- policy and decision-making;
- executive;
- research;
- education;
- extension.

Policy and decision-making

Matters of national soils policy will normally be decided at relatively high government levels, within the appropriate ministry or government-sponsored organization. Once a national soils policy has been agreed upon and adopted, it should act as a guide to all aspects of national development.

Executive aspects

The executive aspects of a national soils policy are mainly related to the technical elements discussed above. They will include:

- The carrying out of soil surveys and recording the monitoring of land-use and farming systems;
- The planning and implementation of soil conservation practices and works;
- The rehabilitation of degraded land and land reclamation.

Research

Research on subjects related to a national soils policy covers a wide spectrum and may be partly governmental at state-financed research institutes and stations, partly at universities or partly private and commercial. National organizations in the developing countries normally receive help from the international and regional organizations. Research activities may sometimes be the subject of bilateral aid agreements.

Education and training

Education in the context of a national soils policy includes both formal education in agriculture, soil science and other subjects at various levels up to that of the university, as well as in more general aspects. In order to create public awareness of the need for soil conservation and related matters and of the measures that can be taken to prevent soil loss and degradation, general community education projects are necessary. Education and research are often carried out together, as in the case of research work for a higher degree at a university. Cooperation between universities and research organizations is thus important.

Extension

To be effective, extension workers must have access to the latest relevant research findings. Consequently, good liaison between research and extension services is essential. An effective extension service is seen as a vital important element in achieving soil conservation on the land itself by transferring technology through advice and assistance to the farmer or other land users. Where economically feasible, a national soil conservation service should be established with an extension service as one of its principal parts.

Other relevant institutional functions

There are other aspects of a national soils policy that require the support of appropriate institutions. These include the provision of finance and credit and the provision and marketing of necessary supplies and equipment. Credit facilities to farmers may take the form of cash or of agreed inputs such as seed, fertilizer, construction materials or equipment.

The aims of national soil legislation

Legislation relating to land and its use at the national level must be related to an agreed soils policy and regarded as one of the important means for carrying out such a policy. Satisfactory legislation depends on the prior definition and adoption of an appropriate soils policy, and the success or failure of that legislation must be measured by the extent to which it does or does not succeed in helping to fulfil the objectives of the soils policy. Soils policies must obviously vary very considerably from country to country according to local environmental, economic and social factors. The various sectors concerned and the pertinent laws should be viewed within the framework of land-use planning and development in order to ensure the harmonious allocation and development of land and soils for agriculture, forestry, recreation, and urban and industrial development. In turn, land-use planning based on an agreed soils policy with a sustained long-term viewpoint will be integrated with the over-all use of all natural resources, taking into account population growth and national needs and aspirations.

Some comments on existing legislation

A common feature of existing land-use legislation in many countries is that it is sectoral in character, dealing with specific aspects of land management such as soil conservation, flood control and land development. In many parts of the world the ownership, distribution, development and protection of the soil are affected by both traditional and more recently developed facets of human society, particularly by legal and para-legal aspects. Traditional concepts of land ownership, land acquisition and rights of use affect how the land is used. In some cases, this results in harmful or abusive use. One such example is temporary ownership, which often results in short-term exploitation of the land; the farmer has little or no incentive to invest in the land and improve it on a long-term basis. The rational development of land is influenced in some countries by traditional social and economic restrictions, such as tribal or feudal systems that influence resource utilization, or by religious restrictions, beliefs or taboos that prevent rational development and use.

Desirable features of national land-use legislation

Land-use legislation is the means through which an agreed soils policy can be implemented. The acceptance of many legislative measures is made easier when their need is explained to people in an educational programme. Conversely, it is very difficult to apply legislation that farmers or others feel is against their best interest. In such cases, farmers have to be educated by demonstrations and other means so that they realize that the legislation imposed on them is in fact designed for the long-term benefit of the community as a whole.

Similarly, creating or attempting to preserve forest areas in places of population pressure and land shortages is of little use unless the state has at its disposal a sufficient force of forest personnel able to maintain the boundaries and keep intruders out. Legislation that fails to take into account practical considerations such as these is unrealistic and likely to be ineffective.

Adequate legislation at the national level will depend very much on local needs, the local physical environment and local social and economic conditions. In general, however, adequate legislation will often include many of the following:

- A clear statement of policies and purposes to explain the need for, and anticipated benefits from, the legislation;
- The establishment of priorities and guidelines in relation to agriculture, urban development, industrial development and other uses aimed at preventing conflicting policies and uses, especially where agricultural production is at issue;
- The power to declare protected zones or areas in which permitted activities are defined, as well as measures to control and prevent waste, over-exploitation, or misuse;
- The creation of institutions and authorities responsible for land resource policy-making and implementation at any required level, and their coordination with institutions and authorities concerned with sectoral aspects such as water, forestry and urban development;
- Adequate powers of implementation and enforcement; these powers will include the traditional ones (a police force and appropriate mechanisms for imposing penalties and settling disputes) but will also include those covering provision of financial contributions or subsidies, credit facilities and the development of a fiscal policy that, through taxation or tax exemption, serves to implement the legislation.

UNEP/FAO ACTIVITIES TOWARDS PREPARATION OF A NATIONAL SOILS POLICY

1. Establishment of formal dialogue with the recipient country with a view to securing signed agreement with the government and the selection of a counterpart organization with which to undertake the necessary research;

2. Recruitment of a multidisciplinary team of experts for every country, consisting of two to three international experts and three to four local counterpart professionals to cover the administrative, legal and technical aspects of soil resource management and land-use policies;

3. Fielding of the multidisciplinary team of experts to the country for a period of up to two months and the preparation of a draft report on the main problems of soil and land use and the relationships to specific environmental conditions and socio-economic needs;

4. Submission of the draft report to the government for review, consultations on the modalities of the adoption of the national soils policies and their inclusion in formulating national development plans;

5. Finalization, publication and presentation of the final report to the government for adoption;

6. Visits to the country to advise on the implementation aspects of the national soils policies and assistance in soliciting donor support for implementation.

FEATURES OF A DRAFT NATIONAL SOILS POLICY DEVELOPMENT BY UNEP/FAO

The National Soils Policy is based on a two months' interdisciplinary mission in the country, including extensive field trips, and is described in five steps or "chapters". After a general introduction in "chapter" 1, the principles of National Soils Policies are described in detail in "chapter" 2, referring to the broad range of issues involved. In "chapter" 3, country-specific issues are dealt with, divided into land-based problems and institutional and legal constraints. The causes and dangers of the various aspects of these problems are described and possible solutions listed. "Chapter" 4 elaborates on the actual proposed rational land use strategies, focusing on three areas:

- improving land use by restructuring agricultural strategies, reforming land tenure, and creation of a flexible legal framework;
- encouraging public participation in land/soil conservation by enhancing awareness, short-term benefits to farmers, establishment of land-users' organizations etc.; and
- strengthening/establishing national institutions to promote/implement the soils policy by creating institutions the country lacks, such as land-use planning bureaus, and soil research organizations; and by identifying workforce, training and research needs.

In "Chapter" 5 specific recommendations for follow-up action are listed. These are largely project ideas of an internationally funded nature.

DISCUSSION

Question: What are your views on converting wetlands into agricultural land?
Answer: The comparative options for the use of this type of land require a great deal of study. There is no general answer.

Question: What are FAO and UNEP doing about the problem of human and animal wastes?
Answer: A great deal of research is being carried out on the utilization of human wastes. As they contain a great deal of heavy metals, it is necessary to treat them for the removal of these elements. In China, human and animal wastes are used to produce biogas. The remaining sludge is used as feed for fish, earthworms and other animals of this type.

Question: Is there any monitoring system that addresses the progress of soil conservation in developing countries?
Answer: In 1987, INFOTERRA carried out a survey that found that some 80 per cent of developing countries had no soils policies. From that point, UNEP began developing a priority list to assist such countries in this area.

Chapter Eleven

The conservation of biological diversity

MONA BJORKLUND
Senior Programme Officer for Wildlife and Protected Areas, Terrestrial Ecosystems Branch, UNEP

INTRODUCTION

UNEP's activities for the conservation of biological diversity have been so numerous and far-reaching that, at the very least, a slim volume would be needed to summarize their most salient features. Let us therefore concentrate on the major milestones that mark this continuum of efforts to protect and conserve the almost unimaginable variety, as yet unknown, of Earth's biotic wealth.

A fundamental landmark in this area of environmental concern is the World Charter for Nature, which was prepared by UNEP, and the World Conservation Union (IUCN) and adopted by the United Nations General Assembly in 1982. It is significant that the initial proposal for the Charter was made by Zaire, one of the world's least-developed countries, rather than by any of the industrialized nations whose citizens were largely responsible for creating the global environmental awareness of the 1960s that led to the Stockholm Conference. The fourth general Principle of the Charter perhaps expresses best that delicate and complex balance essential to the preservation and conservation of what is left to us of the planet's biological treasure-house. That Principle proclaims that "ecosystems and organisms, as well as the land, marine and atmospheric resources that are utilized by man, shall be managed to achieve and maintain optimum sustainable productivity, but not in such a way as to endanger the integrity of those other ecosystems and species with which they co-exist". The Charter concluded by stating that the five general principles it sets forth represent the standard "by which all human conduct affecting nature is to be guided and judged". The moral and practical imperative of this standard cannot be over-emphasized—if only because we ourselves are an integral element of the biosphere.

The World Charter for Nature had been preceded by the first World Conservation Strategy, published and launched jointly in 1980 by IUCN, the World Wide Fund for Nature (WWF) and UNEP. A long, complex document, it is devoted to living resource conservation and divided into three major sections: the objectives of conservation and the requirements for their achievement; priorities for national action; and priorities for international action. However, such have been the discoveries and developments of the 1980s that a second strategy of this kind was deemed essential for the 1990s. Published in late 1991 under the title Caring for the Earth: a strategy for sustainable living, and alluded to in the first chapter of this volume, it is both an analysis and a plan of action. It falls into three parts, the first of which defines the principles of a sustainable society, ranging from respect and care for the entire community of life through enabling communities to care for their own environments, providing a basis for national frameworks for integrating development and conservation and, finally, creating a global alliance. The second part describes 62 actions required for the application of these principles, covering energy, business, industry and commerce, human settlements, farm and rangelands, forests, fresh waters, and oceans and coastal areas. The third and final part of this new strategy of care deals with implementation and follow-up. A fourth milestone is an elaboration of Caring for the Earth, the Global Biodiversity Strategy, launched in 1992 at the World Conference on National Parks and Protected Areas. This, too, is the product of the long-standing partnership of IUCN, WWF and UNEP. Finally, in support of these strategies, UNEP's World Conservation Monitoring Centre has compiled the Global Biodiversity Report, the first comprehensive survey of the states and use of Earth's living wealth. These are all documents. Let us look at some of the ways in which they are put into practice or, in the case of the Global Biodiversity Report, reflect practice.

PROTECTED AREAS

Earlier in this book (Chapter 2, p. 23), Livingstone Dangana of UNESCO's Regional Office for Science and Technology in Africa, outlined UNESCO's Man and the Biosphere Programme (MAB) for protected areas, in whose implementation UNEP has always closely collaborated. Suffice it to say here that there are now more than 300 biosphere reserves in 75 countries representing more than two-thirds of all our planet's terrestrial ecosystems. MAB is devoted not only to conservation, but to interdisciplinary research, demonstration and training.

A significant outgrowth of MAB was the 1983 Action Plan for Biosphere Research, a global review of all such research, including the management of protected areas. One result of the review was a series of courses for biosphere managers in Anglophone Africa, whose evaluation resulted in the publication of a training manual for such managers throughout the whole of sub-Saharan Africa. In addition, on a pilot scale, UNEP is assessing biosphere research in the Russian Federation, East and South-east Asia and the Americas, examining the extent to which these programmes' structures and activities conform to research needs so as to establish standards as well as to develop a methodology for an assessment of the entire protected areas network. One new and important element of this Action Plan is its emphasis on ecotones, the transitional zones between two distinct ecosystems. In cooperation with UNESCO and the Scientific Committee on the Problems of the Environment of the International Council of Scientific Unions (SCOPE/ICSU), UNEP is surveying the quality of these critical areas to improve their management.

Protected Area work is hardly confined to scientists. An ongoing, wide-ranging UNEP/FAO project in the Latin American and Caribbean region attempts not only to take into account the experience of indigenous peoples in the management of the ecosystems in which they live but to train them in various aspects of ecosystem research and management. Another project, undertaken in Panama with the Smithsonian Tropical Research Institute, also involving local people, is an effort to assess the biodiversity and management of the tropical forest canopy. This is particularly important, the more so in view of the damage and destruction of the world's tropical forests because the biotic wealth of such forests resides not in their soils but in their branches and foliage. An additional and dramatic example of the implementation of the World Conservation Strategy covering not only protected areas, but wetlands, fisheries and many other ecosystems that do not enjoy that status, took place in Uganda in 1987, where UNEP and the country's government worked together to produce a comprehensive overview of that war-ravaged nation's natural resources. The result was the publication of a 10-volume series on strategic resource planning and management for the restoration of Uganda's natural heritage.

ENDANGERED SPECIES

How many of Earth's uncounted and unidentified species are endangered? We do not and, for the foreseeable future, cannot know. In view of the massive over-exploitation of wildlife for international trade, UNEP has supported the Secretariat of the Convention on International Trade in Endangered Species of Wild Fauna and Flora (CITES) since that instrument entered into force in 1975. The *Identification Manual for Endangered Species*, prepared and published by the UNEP/CITES Secretariat is a basic tool of customs officers the world over.

Similarly, UNEP supports the Secretariat of the Convention on the Conservation of Migratory Species of Wild Animals (CMS), which has concluded a number of agreements for the protection of endangered migratory species ranging from the European bat to small cetaceans. Moreover, because so many species of waterfowl are migratory, this Convention is of capital importance to wetland and waterfowl conservation under the Convention on Wetlands of International Importance Especially Waterfowl Habitat (the Ramsar Convention), the world's first international conservation treaty. The implementation of that treaty has involved designating some 383 sites covering approximately 27 million hectares of marshes, river deltas and lakes as having international significance in countries as diverse as Argentina, Botswana and Greece.

Last, but hardly least in this brief and highly selective overview, UNEP has been working for many years for the conservation of the African elephant. One of the most significant steps in this effort was the GRID creation of mathematical models of elephant densities throughout the continent. Another more recent activity was the 1992 Meeting of African Range States and Donors for the Conservation of the African Elephant, which reviewed national action plans and national financing needs.

THE ECOSYSTEMS CONSERVATION GROUP

Little of the work sketched above would have been possible without the Ecosystems Conservation Group (ECG), established in 1975, which comprises FAO, UNDP, UNEP, UNESCO, IUCN and WWF and which meets annually to examine the implementation of the World Conservation Strategies and the formulation and implementation of national conservation strategies. It has not only helped to reduce duplication in efforts to deal with the problems of nature and natural resources but has helped ensure a system-wide response throughout the United Nations' family, as well as concerned IGO and NGOs, to this crucial area of environmental action.

DISCUSSION

Question: Must every country develop a national conservation strategy?
Answer: Not necessarily, although approximately 50 have done so. Currently, the best approach is probably that of developing a National Environmental Action Plan, the content of which has been discussed earlier, of which a national conservation strategy would constitute one important element and to which other action programmes, such as a tropical forestry action plan or a national plan to combat desertification, would also contribute. Some funding for the elaboration of such a comprehensive plan is provided by the World Bank. Moreover, *Caring for the Earth* provides a useful guide to the preparation of such plans.

Question: How effective have the older conservation conventions been, notably in comparison with the new Convention on Biological Diversity, signed in Rio?
Answer: Unless a country demonstrates the political will to conserve and manage its natural resources properly—and inspires its people at the grass-roots level to participate actively in so doing—no legal instrument can serve the purposes for which it was elaborated, no matter how comprehensive it may be or how many states have ratified it.

Question: Generally, it is the less-developed countries that are richest in biodiversity and the developed nations that want access to this wealth, including its genetic resources. To preserve this biotic wealth and to make it accessible for development, the developing countries require the transfer of a great deal of technology that the private companies of the developed nations are reluctant to make available. How can a compromise be reached between these divergent positions?
Answer: That is precisely a major aim of the new Convention on Biological Diversity—mutual benefit for both the Northern and Southern worlds.

Question: How were the Protected Areas selected and how can a comprehensive survey of biotic wealth be carried out?

Answer: Unfortunately, most of the Protected Areas were not selected on the basis of even the most fundamental survey. The vast majority were set aside on land considered marginal for developmental uses—in short, on the basis of wholly unscientific criteria. A massive investment of funds and skilled personnel was and is increasingly needed for a thorough inventory of Earth's biotic wealth. The fact that we don't even know how many species exist demonstrates our negligence of this subject. Moreover, we will never be able to calculate that number because so many species will have become extinct before such a calculation is possible. This is an irremediable loss for our knowledge of the dynamics of the biosphere because we must know the ecological role of each species, as well as its economic value, if we are ever to achieve sustainable development.

Chapter Twelve

Genetic resources

HAMDALLA ZEDAN
Senior Programme Officer for Biological Diversity, Microbial Resources and Related Biotechnologies, UNEP

Intimately bound up with biological diversity—which is simply all animals, plants and microorganisms, as well as the ecosystems in which they live—is the *viability* of life itself. Most estimates place the number of species to which Earth plays host at 5–50 million. However, there may well be 80–100 million. And of these, to date, only 1.4 million have been identified, let alone described. Because of human pressures alone, some 100–150 of these species are being destroyed *each day*— most of them through habitat destruction and habitat fragmentation. Pollution also plays a major role, as does virtually any element that upsets the balance of a given ecosystem. The forthcoming global warming will doubtless destroy many more.

The distribution of species varies widely globally. In general, they are more sparse at the poles and tend to increase as one nears the Equator. Their numbers reach their greatest heights in tropical forests and their marine equivalent, coral reefs. And each member of each species, from a human being to the smallest microorganism, is an *individual*. Each also performs the same basic physiological functions: mammals with many organs, a microorganism often with the components of its single cell.

These millions of species have evolved over 500 million years, developing under widely different environmental conditions and adapting to changes in these conditions, notably those of temperature. Thus, over the eons, though countless groups have become extinct, the number of species has proliferated. Their adaptation has usually taken the form of mutation, i.e. some change in their genetic material.

From the earliest days of human development, men and women have recognized the value of a number of species to fulfil their basic needs. To this end, shortly thereafter, they perceived the value of specific populations, even of individuals. These they bred and, in many cases, cross-bred. This, in turn, increased the range of species. It not only extended the diversity *between* species, but *within* a given species. In short, just as biotechnologists do today, early humans deliberately altered the genetic makeup of plants and animals they found useful—with one significant difference: no one tried to manipulate chromosomes or select just one or a few genes from an entire band. In other words, they did not narrow the genetic base.

Fundamentally, genetic resources may be defined as any and all species of actual or potential value for socio-economic development in any area. All these organisms provide goods and services which can be categorised under two headings, production and consumption. If one or another genetic resource is tradeable on the market, it is assigned a value there. Others are consumed or otherwise used by human beings without passing through the market; a variety of forest resources, for example, are used by many indigenous peoples without passing through any market except, perhaps, some simple form of barter. Consequently, they bear no price tag. And if a resource is not traded in some way, it is not valued as "productive". Hence the enormous problems in assigning economic values to genetic resources, notably that host of unknown species, any one of which may some day help cure a disease, augment our food resources or perform any one of a vast spectrum of functions that we cannot begin to imagine.

Only recently has information begun to accumulate on the value of genetic resources, especially those involved in agriculture. From their earliest agricultural endeavours, human beings have been very selective in choosing from nature's gifts. Although some 80,000 edible plants exist, only some 150 are cultivated and only 20 species account for 90 per cent of the world's food production. The "Green Revolution" vastly heightened the productivity of a number of the world's major food crops. However, the hybrids that were developed are far more vulnerable to environmental stress, especially pests and diseases. This is due to the fact that the genes that protect a given plant are very close on the chromosome they share to the genes that increase its yield. Resistance was sacrificed to productivity. And the only way to restore that resistance—or to endow the cultivar with other desired traits—is to breed the plant with wild strains.

We can now begin to assign figures to some genetic resources. In Asia in the late 1970s, several wild rice and wheat strains were used to improve the resistance of existing crops. The result was an increase of US $1.5 billion annually in rice sales and US $2 billion in the sales of wheat. More recently, one gene from a strain of Ethiopian wild barley was used to save California barley from a virus; profits rose by US $160 million annually in the United States alone. Consequently, the value of all wild agricultural species is currently set at some US $100 billion annually—and that is very probably an underestimation.

Added to these figures are benefits we have not yet even begun to quantify. Crops that have greater resistance will need fewer pesticides—and fewer pesticides, in the end, will mean less toxicity and less pollution. Wild strains can also be used to develop crops that can be grown on marginal land—or on water.

To turn to the field of medicine, one of every four pharmaceuticals contains active ingredients taken from wild plants whose cultivation or synthesis in laboratories would be prohibitively expensive. More than 300 antibiotics are produced from microorganisms found in the wild. During the 1960s, only one leukemia-stricken child in five could hope to survive. Now, because of genetic material extracted from the tiny wild rosy periwinkle, which grows in the forests of Madagascar, four in five will live. The sales of this one drug alone total more than

$150 million annually. It should be noted, too, that given the efficacy of a number of traditional cures that have emerged in recent years, Western pharmaceutical producers are consulting healers among indigenous peoples the world over and exploring their lands.

The conservation of genetic resources must be continued both *ex-situ* and *in-situ* in biosphere reserves. The latter method is preferable by far. One never knows what infinitely valuable organism lives in what relationship with the precious one that has been found in the same ecosystem.

DISCUSSION

Question: Does climate change necessarily result in the destruction of biological diversity?

Answer: There are five cases of mass extinction recorded in connection with climate change. The last one, 65 million years ago, was caused by changes in temperature. It must be remembered that unlike most animals, the vast majority of plants cannot move; their capacity for adaptation is therefore limited to a great degree.

Comment: Given the profits made by transnational corporations, there is great need for the transfer of technology to help developing countries to make use of their genetic resources.

Comment: The protection of biodiversity is very important. It should be carried out by governments rather than left to the play of market forces.

Chapter Thirteen

Depletion of the ozone layer

PETER USHER
Coordinator for Climate, GEMS Programme Activity Centre, UNEP

In the stratosphere, some 20–50 km above Earth's sea level, lies the thin—and thinning—layer of ozone whose depletion so preoccupies the world today. Ozone is a poisonous gas. It is also a greenhouse gas, like carbon dioxide; when present in the troposphere, it contributes to global warming. In the stratosphere, however, it has an enormously redeeming quality; it shields us from the most dangerous form of ultraviolet radiation, UV-B, one of whose minor effects is sunburn, but which, unchecked, can ravage virtually every form of life—stunting vital crops, disrupting the marine food chain and causing blindness and skin cancer. A more complete list of the effects of ozone depletion is given below:

- Increased UV-B radiation reaching earth's surface;
- accumulation of tropospheric ozone and acid aerosols, causing worsening air pollution and acid rain;
- photochemical formation of tropospheric ozone;
- damage to materials: paint, plastic, rubber;
- damage to biological links in human food chain;
- crop and forest damage;
- suppression of body immunity: increase in infectious diseases, less-effective vaccination;
- eye damage: increased incidence of cataracts and blindness;
- aquatic life: less ocean plankton, lower fish harvests;
- other human health effects: respiratory illnesses and heart problems;
- rise in incidence of skin cancer;
- terrestrial ecosystems: reduced crop yields, stunted plant growth.

Serious concern about ozone depletion emerged in the late 1960s and early 1970s, as major annual reductions of the gas were increasingly observed over the Antarctic during the southern hemisphere spring (i.e. September/October). Losses had been detected before, but had been ascribed to natural stratospheric breakdowns of the gas interacting with other chemicals from the troposphere in the presence of sunlight. However, the magnitude of the Antarctic hole spurred the undertaking of an industry-sponsored research programme in 1972. Two years later, Sherwood Rowland and Mario Molina of the University of California attributed the breakdown of stratosphere ozone specifically to chlorofluorocarbons—or, as they are better known, CFCs. The complex stratosphere chemistry of this process starts when chlorine reacts with ozone, breaking it down to form oxygen and chlorine monoxide; the latter reacts with free oxygen to reform chlorine and the cycle starts again. When nitrogen oxide is present, the cycle is disrupted, long-lived chlorine compounds are formed and the rate of ozone depletion slows down. However, on polar stratospheric clouds, long-lived chlorine compounds are converted to highly reactive forms of chlorine that destroy ozone.

What are CFCs? They do not occur in nature. They are wholly man-made combinations of carbon, hydrogen, fluorine and chlorine. Stable, non-toxic, non-corrosive and inexpensive, they seemed ideal for a host of domestic and industrial uses. They constitute, for example, the staple of insulators for building construction, the stuff of mattresses and other furniture; they are also used in dry-cleaning agents, propellants and packaging materials. An analogous class of chemicals containing bromine rather than chlorine, known as halons and used largely as fire extinguishers, have a similar, albeit lesser effect on the ozone layer.

When Rowland and Molina propounded their theory in 1974, no one could confirm that the process they had discovered in the laboratory was indeed taking place in the stratosphere. None the less, CFC use declined, notably in the United States, which acted almost immediately to ban these compounds in propellants, and there was time to test the theory and to establish a process whereby risk could be assessed. It was at this point that UNEP stepped in, setting up the Coordination Committee on the Ozone Layer, a group of scientists named by governments of countries that had active research programmes on ozone depletion. Under UNEP auspices, the Committee made regular assessments of the state of the ozone layer and predictions of what might happen to it if the use of CFCs increased.

And it did. By the late 1970s, many more applications were found for CFCs and production again began to rise. It fell off somewhat again in the early 1980s and little ozone depletion was observed, apart from natural fluctuations—seasonal changes that differed little from those observed during the 1950s and even variations of as much as 25 per cent within 24-hour periods between cities such as Miami and Washington. How could one determine the extent of loss due to the decomposition of anthropogenic chemicals when natural variations marred the signal? We suspected only that if there was little or no growth in CFC production and consumption, there would be little or no long-term ozone depletion. Nevertheless, the need for international action was so strongly felt that by 1981, in accordance with a decision of the UNEP Governing Council, a working group of government, legal and technical experts began drafting the Vienna Convention for the Protection of the Ozone Layer, a framework legal instrument that bound its signatories to cooperation in research, exchange of information and monitoring of the ozone layer. However, the text said nothing about how CFCs should be reduced.

The Convention came into force in 1985, the very year in which a massive ozone hole was discovered by satellite over the South Pole. By 1987, half the ozone of the area had disappeared. However, according to some of the theories of the time, this loss was not necessarily due to CFCs; several ascribed it to nuclear testing, others to volcanic activity that propels enormous amounts of ozone-depleting materials into the atmosphere. However, because we knew at least the impacts of ozone depletion and were well aware that, although an unpopulated area, Antarctica is the cradle of fishery productivity for the entire southern hemisphere, there was an enormous burst of research activity that pointed increasingly to the central role of CFCs.

Despite the fact that some nations had begun to curtail CFC production and use unilaterally, a deep need was felt for developing an international process for the reduction of these compounds. It took only two years to finalize the Montreal Protocol on Substances that Deplete the Ozone Layer, which was approved in 1987 and which came into force in 1989. Ratified by 34 countries that accounted for some 80 per cent of CFC consumption, the Protocol called for diminishing the production and use of the five major CFCs and the three major halons severely by 1998. Special provisions were made for developing countries with low per capita incomes.

The Montreal Protocol was the first pre-emptive international agreement on the global environment—and luckily so; by 1989, it was confirmed that an Arctic potential for ozone destruction comparable to that of Antarctica was probable. Further, that depletion might well spread to heavily populated areas of the world, seriously affecting the populations of northern Europe, Canada, Greenland and Siberia.

Only by 1990 was there clear proof that CFCs and halons were responsible for ozone depletion of the magnitude observed. At the same time, there were encouraging prospects on the technological front for the production of inexpensive substitutes for a wide variety of CFC uses. Consequently, in 1990, it was possible to tighten the Montreal Protocol stringently and to anticipate a complete halt of ozone-depleting substances by the year 2000—and again fortunately. Were the Protocol to have remained only as strict as it was in 1987, the Antarctic ozone hole would probably never disappear.

Our problems now are twofold. The first is technological. The short-term phase-out of CFCs depends on continuing to use hydrochlorofluorocarbons (HCFCs) which, although they destroy ozone, do so to a much lesser extent than CFCs and survive intact in the atmosphere for a much shorter time. From then on, say 1994–1998, how long will it take to replace the controlled substances with hydrofluorocarbons (HFCs), which are ozone-friendly because they break up in the troposphere (although they, too, contribute to global warming)? HFCs or blends of such chemicals are now available, but the upgrading and replacement of existing equipment will take some time. We are talking here only about refrigerants, probably the most difficult of all the technological problems. However, industry maintains that it must continue using CFCs and HCFCs so as to carry our further research and development on HFCs. And one cannot overemphasize the fact that though our knowledge of ozone-friendly chemistry has increased far more rapidly than anticipated, we still do not know enough.

The second major problem is of course political: *obtaining full compliance* with the amended Montreal Protocol and whatever revisions that may be incorporated into it as a result of the four-year assessments built into its original text. Theoretically, we could eliminate the emissions of ozone-depleting compounds tomorrow. Even then, we would have to wait for perhaps as much as a century for natural repair of the ozone loss that we know exists today.

DISCUSSION

Question: Is there any way of replacing the lost ozone in the stratosphere?

Answer: There have been several notions of replacing the gas by sending rocket-propelled loads up into the stratosphere, but this would probably do yet more harm to the ozone layer. In addition, some scientists have spoken of the idea of destroying the chlorine in the stratosphere by laser rays. But the real answer is probably not a *technofix*. It is probably better to stop emitting harmful substances into the atmosphere—allowing it to repair itself rather than risk the emergence of a new environmental problem that a technofix might create.

Chapter Fourteen

Freshwater resources: their depletion, contamination and management

HABIB EL HABR
Programme Officer, Terrestrial Ecosystems Branch, UNEP

INTRODUCTION

Ever since our emergence on Earth, human beings have been contaminating that most precious of nature's resources—water. As our numbers have multiplied along with the scale and complexity of our activities, notably in agriculture and industry, we have imposed greater and greater demands on this finite, though renewable, resource. By the year 2000, we may well need two to three times as much freshwater as we used in 1980. Such projected demands appear likely to exceed the potential global water supply for the foreseeable future and beyond. Moreover, in our efforts to attain an economic growth that will meet the needs of the world's population, we have indulged in development at virtually any cost—eroded and salinized soils, damaged and destroyed forests, seen an increased incidence of environment-related diseases and, of course, polluted waters. No country, developed or developing, has escaped these problems.

As we increasingly realize, this kind of development undermines our resource base for further progress. The only road open to us is sustainable development. At this time, however, there is no accepted definition of sustainable development beyond, perhaps, the simplest one provided by the World Commission on Environment and Development, i.e. development that meets the needs of the present without compromising the ability of future generations to meet their own needs. Further, though, the Commission noted that sustainable development is not "a fixed state of harmony, but rather a process of change in which the exploitation of resources, the direction of investments, the orientation of technological development and institutional change are made consistent with future as well as present needs". Let us examine the applicability of this concept to the sustainable management of freshwater resources.

THE RESOURCE-ORIENTED APPROACH

Traditionally, freshwater management has been carried out within a resource-oriented approach. This means that ever more resources were exploited without commensurate regard for the conservation of water, environmental protection and a number of other considerations. Usually, when demand exceeded resources during low-water periods, dams were built to regulate the water supply in terms of time. When, despite such regulation, demand exceeded supply, projects were developed to exploit waters from outside sources. In addition, in the general process of water development, single-purpose projects were replaced by multi-purpose projects.

Usually, the resource-oriented approach entails a variety of mounting undesirable side effects such as the deterioration of the environment, decrease in the income of the population and changes in its social structure. Even when the design of water projects includes an assessment of possible side effects, the evaluation of the project's efficiency is often incorrect either because the environmental impact assessment was not properly carried out or mitigation measures have not been implemented. Moreover, the factors encompassed by the cost-benefit analysis are often selected arbitrarily. For example, in planning its network of water reservoirs, one country used a methodology of cost-benefit analysis in which the only expensive factor was the cost of transferring the settlements that were to be inundated; the considerable losses of valuable agricultural land were simply overlooked. The question of beneficiaries also calls for careful consideration. When too much of the energy generated by hydropower projects is channelled to urban populations, farmers are often displaced by the water reservoirs created by those massive undertakings or their production cycles are disrupted by the changes in the new hydrological regime. Consequently, both urban and rural peoples lose because food production is diminished.

We must, however, realize that despite the criticisms of the negative effects of many hydropower projects, significant benefits result from properly designed and exploited ventures of this nature, as well as many other types of water developments. One clear example is the Aswan High Dam of the Nile, which has engendered so much negative comment that its obviously positive effects have been slighted. Much of northeast Africa has experienced a low-water period during the last few years. Without Lake Nasser, created by the High Dam, irrigation in Egypt would have suffered enormously, as would the energy supply of that country.

Fortunately, the resource-oriented approach to water management is changing as this resource becomes increasingly scarce. We are witnessing more and more attempts to increase the efficiency of water use. Price for water is being introduced or increased. Such processes lead further to mechanisms that match supply with demand. Where a real need exists to adjust demand, it usually occurs where the economy of the water basin is already quite complex and well developed. In such cases, different sectors of the economy have different water demand patterns that exert strong influences on water resources, both quantitatively and qualitatively. Many of these demands are competitive, one activity of one sector in one area of a basin conflicting with another. Consequently, the quantity of water may be insufficient and its quality may be far below the standard desired. Moreover, further development of the basin may be impeded. In short, the development of the basin may well become unsustainable.

SUSTAINABLE DEVELOPMENT OF FRESHWATER RESOURCES

To address the goal of the environmentally sound management of freshwater resources, we must consider a variety of functions that natural waters perform. Their necessity to human life is obviously essential. Underlying this function and reaching far beyond it is the role played by water in all types of ecosystems— natural, modified and anthropogenic. Water is a major factor unifying ecosystems because the exchanges of matter and energy pass, to a great extent, through the water cycle. Any changes in a given ecosystem lead to changes in parameters, both quantitative and qualitative, of the water components of the system. Therefore the role of water in ecosystems can be compared with that of blood in an animal's body. Unfortunately, this role, although well recognized by scientists, is frequently ignored in the management of territories. Yet another closely related function of water is that of a carrier of materials in global biogeochemical cycles. Thus water as a component of ecosystems may well be considered as a local branch of the global cycle.

This leads us to UNEP's programme of environmentally sound management of the inland water resources of entire water systems, known by its acronym EMINWA. The central aim of this programme is the introduction of a comprehensive approach to the planning and management of freshwater resources on a basin-wide scale. The major goals of the programme are the following:

1. to assist governments to develop and implement environmentally sound water management programmes for inland water systems and to use this approach for demonstration purposes elsewhere;

2. to train experts and establish training networks in developing countries to implement environmentally sound water programmes, including drinking water supply and sanitation programmes;

3. to prepare a manual of principles and guidelines for the environmentally sound management of inland water systems;

4. to make regular world-wide assessments of the state of the environment for such systems; and

5. to inform the mass media on the activities and achievements of this programme and to increase public awareness of environmentally sound water development.

Of these five goals, the highest priority is encouraging countries that share a common river, lake or aquifer basin to develop their water resources sustainably and to utilize them without conflict. This element of the EMINWA programme has much in common with the UNEP Regional Seas Programme, which has succeeded in creating cooperation among countries bordering a common sea, including some nations that were or are actually belligerants. Although each inland basin must be approached differently, the following three phases are involved in each regional programme:

1. organizing of a working group of experts designated by the appropriate ministry of each basin country to conduct a diagnostic study of the current status of water management and the environmental problems of the basin; the objectives of this study are to define specific environmental problems and their impacts for the present and the foreseeable future and to assist the basin governments in formulating programmes for the incorporation of environmental concerns in their management of water resources;

2. strengthening the awareness of the various governmental institutions involved in socio-economic development activities about their potential impacts within the basin; and

3. encouraging possible donors to contribute to the implementation of water development projects within the basin.

The diagnostic study provides the basis for the preparation of an action plan to overcome the problems identified and to promote the environmentally sound management of the entire basin as part of the strategy for the sustainable development of the entire region. To implement the action plan, a variety of legal and institutional arrangements must be made. Typically, this includes a convention or agreement expressing the willingness of the governments involved to cooperate in carrying out the action plan. Such agreements also specify the modalities of the plan's implementation. Finally, the plan is put into effect through a series of projects financed usually through a combination of the basin countries, United Nations' bodies, the regional organization and donor governments.

The EMINWA programme is designed to deal with river and lake basins, as well as groundwater aquifers, priority being accorded to international water systems. UNEP's first such experience has been the Zambezi river, 3000 km in length, which drains an area of 1 300 000 square km. Some 20 million people live within that basin, which includes eight countries: Angola, Botswana, Malawi, Mozambique, Namibia, Tanzania, Zambia and Zimbabwe. Fortunately, there have been no significant conflicts in the utilization of this river system and it is hoped that the coordinated implementation of the Agreement on the Action Plan for the Environmentally Sound Management of the Common Zambezi River System, adopted by a conference of plenipotentiaries in 1987 and designated as a programme of the Southern African Development Coordination Conference (SADCC), will avert any possible future conflicts. UNEP's second EMINWA experience has been the preparation of the master plan for the environmentally sound management of the Lake Chad basin area, finalized in August, 1991. At the same time, other regional groups on other continents are exploring the identification of other river, lake and aquifer systems to develop similar action plans and agreements. Among these are the Orinoco River basin, the Aral Sea basin, the Lake Titicaca basin and the Nile River basin.

As indicated earlier, the development of manuals and guidelines constitute another major component of the EMINWA programme. In this particular area, UNEP expects to issue four publications on the following subjects:

1. methodological guidelines for the integrated environmental evaluation of water resources development;

2. a code of conduct on the prevention and management of accidental pollution of transboundary inland waters;

3. guidelines on lake management; and

4. decision support systems for the management of international rivers.

The education and training of skilled personnel is also deemed a prime requirement for EMINWA success. This field will encompass the development of standard training materials, textbooks, curricula and syllabi on the environmentally sound management of water resources. The organization of training courses, seminars and workshops is also a major element of this endeavour. Moreover, the establishment of specialized networks of training institutions is envisaged. In its attempt to realize all these goals, UNEP is working closely with other United Nations' bodies such as the United Nations Development Programme, the World Health Organization and the Food and Agriculture Organization of the United Nations, as well as regional organizations such as the Lake Chad Basin Commission, SADCC and the Organization of American States in addition to numerous regional and national organizations, both governmental and non-governmental.

In conclusion, let us understand fully that environmentally sound, comprehensive, basin-wide water resources planning and management is the only viable contribution of the water sector to a global strategy for sustainable development.

DISCUSSION

Question: Given UNEP's catalytic role, could it not persuade countries with conflicts concerning water resources to work together to develop their shared resources? Could it not suggest some joint project to stimulate cooperation?

Answer: UNEP's role does not extend in any way to interfering between the interests of two countries concerning a shared natural resource. We cannot pressure any government on behalf of another. As to projects, UNEP can act only at the req' st of governments. Their willingness to cooperate is an essential condition fo' iaking any project work.

Chapter Fifteen

The environmental consequences of energy production, energy transmission and energy use

BEATRICE KHAMATI
Consultant, Technology and Environment Branch, UNEP

INTRODUCTION

As we are all aware, energy is central to all human activity, from basic household chores to all forms of industrial production. Indeed, the major current global environmental issue—climate change—is fundamentally an energy usage problem.

Most energy sources, both renewable and non-renewable, are ultimately derived from the sun—fossil fuels and biomass among the most obvious. But if we think of hydropower, we must reflect that the water cycle is regulated by the sun. Similarly, wind energy is created by sun-created temperature differentials that bring about the differing pressures that cause winds. Only the terrestrial energy forms— nuclear and thermal—derive ultimately from radioactive sources in the Earth's core.

FOSSIL FUELS

The major industrial fuels—oil, coal and natural gas—are the most significant polluters of the atmosphere because they are basically composed of carbon and hydrogen and therefore react with the air's oxygen to produce carbon dioxide, the chief greenhouse gas. Coal is the greatest pollutant because it has a high sulphur content and therefore contributes significantly to acid rain. Let us remember, too, that the mining of coal has the highest occupational risks in the whole of the energy sector; that it produces land degradation, partly because of acid mine drainage into soils; and that its extraction, notably for surface coal, often involves deforestation. Coal none the less provides some 27 per cent of the world's energy, approximately 6 per cent in developing countries, although rates vary enormously among these.

At the other end of the pollutant spectrum is natural gas, the cleanest of the fossil fuels. Because of its high hydrogen content, it produces the most energy per unit utilized. It also has the lowest sulphur content. In addition, most natural gas still remains untapped. Though it is usually found close to coal, its extraction involves very different and far more sophisticated techniques. None the less, it is methane, another greenhouse gas, which often leaks or flares into the atmosphere where coal is mined, producing pollution as well as soil contamination. Superficially, shifting from petroleum to natural gas seems to be a viable option for many countries, but its implications call for careful examination.

RENEWABLE SOURCES OF ENERGY

Biomass

Let us first look at biomass, "the fuel of the poor", which constitutes only 1 per cent of energy consumption in the developed world (albeit 4 per cent in the USA), but 44 per cent in developing countries—in some of the poorest countries, almost 90 per cent of energy requirements.

Biomass includes wood and charcoal, crop residues, animal wastes (as well as human wastes, particularly in municipalities) and those energy crops that produce oils and alcohols. In many rural settings, biomass is the stuff of most of the basic physical aspects of life: 10–25 per cent of it goes *into* the pot and 60–70 per cent *under* the pot, while 5–30 per cent is shared between livestock and the land.

Animal wastes, particularly dung, are a poor energy source. Its conversion efficiency is low, with great waste production, much of which, admittedly, becoming fertilizer. Though the biogas from this waste is biodegradable and can be used as an energy source: how much livestock would be needed for a productive biogas plant? Thus would a poor man's resource become that of a rich man. Although there have been a number of installations of communal plants, notably in Kenya, the approach has not worked well because of frequent sociopolitical problems often involving such aspects as accountability and the delegation of responsibilities. A closely associated issue in contemplating such plants is that their utilization reduces the availability of manure to fertilize the land.

By contrast, oil crop potential is under-estimated and under-exploited. For example, the euphorbia species, a hardy plant that thrives in marginal areas and which is used to fix sand dunes at the edges of the desert, is renewable and produces solid, liquid and gaseous fuel. Its sole drawback is that it gives off a bit of carbon dioxide.

The negative impacts of biomass, so far as we know them, are largely local. In and around the cooking area, carbon-monoxide and -dioxide particulate pollution cause eye and respiratory tract infections, the latter a major child-killer. Although much has been said about the role of fuelwood in deforestation, it must be remembered that in many rural areas, the wood available as fuel to the local population is only what is left over after the rest has been sold for other purposes, among them, commercial timber. The linkage of fuelwood use to global warming is still not quite clear, as the stumps and roots of the trees felled for this purpose remain in the earth, carbon sequestered in the soil. Although it is well known that forests act as carbon-dioxide depurators, we do not yet know how much cutting trees reduces this depuration effect.

Consequently, taken as a whole, biomass may well be a fuel for the 21st century and beyond.

Geothermal energy

Geothermal energy is not strictly renewable in the sense that we think of renewables as being replenished at the same rate at which they are harvested. Often, once one geothermal outlet is tapped, all the others around it are depleted in varying degrees. In New Zealand, many hot springs have simply ceased to exist. However, it is possible to make geothermal resources renewable by reinjecting the used steam and hot water into their source.

None the less, geothermal energy does have certain disadvantages. Let us remember that we are not talking of pastoral geysers. Because of its subterranean sources, geothermal steam is often laden with heavy metals, frequently poisonous such as arsenic and mercury, which are discharged into waterways thereby making their way into the food chain. Among these toxic substances, one often finds hydrogen sulphide, a component of acid rain, which has a corrosive effect, as well as ammonia and carbonates which also enter waterways—as does the steam's heat itself, adding another pollutant. Moreover, there is more sulphur generated by a geothermal plant than by a normal coal power station. Even more discouraging, most geothermal plants release more radioactive material, notably radium, than does a stable nuclear plant—perhaps, however, because the radioactive aspect of geothermal energy tends to be ignored. Then, too, the steam erupts from the geothermal source at such high speed that it causes imbalances such as land subsidence and noise pollution. Fortunately, as geothermal plants are generally located far from built-up areas, these impacts do not have long-term effects on their natural surroundings unlike hydro-electric plants, which often require large-scale dam construction and associated projects for flood control.

Solar energy

The low density of solar energy generally calls for large land areas for the construction of solar collectors. It must also be remembered that solar energy is intermittent, variable and unevenly distributed. Yet it is very versatile in its uses, which range from cooking, heating and the drying of various crops through cooling, refrigeration, brewing and the pasteurization of milk. For its use in residential areas, special buildings are needed. In this connection, one should think of sustainable cities surrounded by solar collection fields.

Widespread use of solar energy usually requires transmission fluids, many of which are toxic, such as freons. Another hazard in the use of solar energy is helostasis: where solar collectors are mirrors, fires can be caused if these are not correctly oriented. One proposed scheme for the large-scale use of solar energy is a synchronized series of satellite power transmitters beaming the collected energy towards Earth. It must be noted, though, that this beaming area must be high enough above the planet and its transmitters placed sufficiently far apart so as to avert incineration. Finally, to satisfy Earth's power needs largely with solar energy (supplemented by some wind and hydro power) 7 million square miles of land and ocean would be needed. This presupposes using only desert lands—with or without oil beneath them—otherwise unusable. Even so, so large a scheme might risk altering the Earth's albedo (i.e. the fraction of incident radiation reflected by the Earth's surface) thereby producing climate change.

Wind energy

Wind energy, like solar power, is intermittent and unevenly distributed and has low density (though higher than that of solar sources) but is totally renewable. It may call for solar batteries, which often contain acids harmful to the environment. The use of wind energy would have to be coupled with another energy resource, such as hydropower or a wood or coal-based generator to satisfy the needs of a community.

Hydropower

The use of hydropower usually interferes with the natural flow of a river, except where high natural falls exist. Even where they do, such dynamic strong flows cause siltation, interfering with the agricultural uses of alluvial soil downstream. One prominent occupational hazard of the use of hydropower is the break-up of dams, whose immediate fatality rate is often higher than that of a nuclear disaster. Hydropower also alters the micro-climate of its surroundings, raising the water table and increasing the humidity—and, in so doing, introducing disease vectors such as mosquitoes and bilharzia snails.

Even though hydroelectricity appears to be a "clean" form of energy, it is none the less associated with water pollution, notably thermal pollution, which may disturb or destroy aquatic ecosystems. It should therefore be borne in mind that hydro-electric plants require huge cooling towers if the environment is not to be disturbed. It must also be remembered that during drought or lower-than-normal-rainfall, dams may deplete the surrounding watershed.

LOOKING TOWARDS THE FUTURE

Although renewable sources of energy appear ideal for use in developing countries, given their low-density nature and the low-density population of much of the developing world, we have seen that they have two major drawbacks. First and foremost, they require enormous technological inputs, most of which are not immediately transferable to most developing countries. Second, despite their reputation as clean energy sources, many are not necessarily environmentally beneficent. Indeed, the only "clean" source of energy appears to be nuclear. However, replacing all coal-powered electricity systems with nuclear plants, even in a medium-term growth scenario, would require 5000 large nuclear plants by the year 2020. This would cost approximately US $60 billion. Public electricity spending in *all* the developing countries amounts to less than US $10 billion annually. Who would furnish such financing—apart from the sums needed to dispose of nuclear wastes, to say nothing of the huge dangers of nuclear power, both physical and psychological?

The broad answer, I believe, involves increasing energy efficiency through a vast variety of technological changes, most involving transfer, but much also inhering in the improvement of indigenous technology. By improving vehicle efficiency, for example, the United States alone could liberate 208 million barrels of oil per day, which could be used to double agricultural production in the developing countries. And only 5 per cent increase in oil supply world-wide would eliminate fuelwood consumption in most developing countries.

The UNEP energy programme focuses largely on strengthening national capabilities for energy policy and planning. That subsumes virtually all the above.

DISCUSSION

Question: How can wind energy be stored?
Answer: Once it has been converted into electricity, it may be stored in batteries. A better solution, however, is to couple the wind generator with a generator based on another source such as diesel or hydro energy. In fact, energy storage is very expensive.
Comment: Pump storage systems, such as those in the mountainous area of Wales, can be used to store any kind of energy. The excess energy (or that which is produced only during the day, like solar energy) is used to pump water from a lower reservoir to one higher up the mountain. During the night—or whenever the additional energy is needed, water is allowed to flow from the upper reservoir into the lower one, generating electricity as it moves.

Question: Is there a UNEP regional programme to minimize power consumption and emissions?
Answer: So far, no. UNEP has been concentrating on large individual developing countries such as China, India, Brazil and Mexico.
Comment: Often, as in the case of solar cookers, the technical solution exists, but turns out to be impractical, because it takes so much longer to cook. In order to identify optimal solutions, one must clarify the problem one wants to solve and consider the key importance of local lifestyle and habits.
Comment: The problem of forests in the developing world should not hide the massive and still continuing destruction of forests in the developed countries. Entire civilizations have based their development on forest exploitation, starting with the ancient Romans. Today in some developed countries, specifically the United States, land ownership rights may conflict with government plans to stop deforestation.

Part III

CASE STUDIES ON ENVIRONMENTAL MANAGEMENT

Chapter Sixteen

Environmental dynamism in the promotion of Ghana's cultural heritage: the communication medium

KWASI DANKAMA QUARM
National Commission on Culture, Environmental Protection Council, Ghana

INTRODUCTION

As custodian of Ghana's cultural property, the National Commission on Culture has become one of the official institutions responsible for the conservation of the environment. The Commission is headed by a secretary of state who is responsible to the office of the head of state.

CONSERVATION FUNCTIONS

In the field of conservation, the Commission collaborates with other bodies, such as the Environmental Protection Council, the Department of Game and Wildlife and the Forestry Department, to preserve all unique natural features, sites and scenic areas with recreational, aesthetic, religious, scientific and educational utility to enhance their tourist potential. The Commission not only plays the role of watch-dog but that of educator, information agent and opinion-leader.

THE CONCEPT OF TRADITIONAL CONSERVATION

Traditional wisdom holds the environment sacred. This sacredness is symbolized artistically by statuettes of a woman of tranquil radiant beauty with a distended abdomen and enormous breasts. She signifies the actual and potential ability of the environment to give birth and sustain human beings adequately if they pay her their reciprocal duties of awe and care. She is, after all, pregnant, the bearer of the land's eternal renewal and commands the reverence and esteem due to her burden of continuously bearing the vital gifts of the Earth.

The Commission, recognizing this traditional religious view, uses traditional drama, music, drumming, dance and oratory in contemporary plays, concerts and musical shows and festivals to portray the disastrous consequences that could result from the abuse of the ancient wisdom represented by the goddess in our own modernizing era, in which science and technology have made the future appear bleak.

The articulation of this traditional view of conservation is maintained in practices such as:

1. The use of traditional prayer forms at public and official functions. These serve as reminders of human dependence on the divine and nature for their ultimate survival.

2. The encouragement of the use of the Fertility Goddess in homes and offices to serve as a sort of public proclamation.

3. Encouraging respect for clan totems and taboos. The use of these mechanisms in the past tended to reduce dependence on certain rare animals or plants in a locality.

4. Protection of sacred groves and tribal lands.

RESULTS

The programmes have had tremendous impact on the Ghanaian populace. High government officials are in the vanguard of the conservation crusade. There is a national celebration of Environmental Awareness Day (which, in 1991, fell on Saturday, 16 November, and was marked by a brass-band picnic in the principal streets of Accra, Ghana's capital).

There has been an increase in size and number of non-governmental organizations, among them the Green Forum, Friends of the Earth, the Evergreen Club of Ghana, the Save the Seashore Birds organization, the Wildlife Protection Society and the Wildlife Foundation.

New protected zones have been established, among them the Fiema and Buabeng Wildlife sanctuary in Nkoranza District of Brong Ahafo Region (in the central part of the country in the savanna ecosystem) and the Shai Hills Game Reserve in the Accra Plains.

The Tourism Development Scheme for the Central Region (TODSCER), a World Bank assisted project has also created protected zones in the Central Region. In addition, many communities have participated in tree planting and bush-fire control to arrest desertification. Moreover, standards of environmental cleanliness have risen significantly because of improved systems of waste disposal and drainage.

CONCLUSION

The method may appear anachronistic but its success has been tremendous. The idea of conservation is carried far and wide through the public shows that are often broadcast on the national television and radio network. Centres of National Culture are established in all districts and regions and managed by advisory bodies comprising the chiefs, prominent citizens and heads of government institutions in each particular area. Thus, through the promotion of its cultural heritage, Ghana has been able to create an effective communication medium for the dissemination of environmental education and information. Its example may be useful to other countries.

Chapter Seventeen

Oil pollution in the ROPME sea area

ALI KHALIFA EL-ZAYANI
Environmental Protection Committee, Bahrain

INTRODUCTION

The ROPME sea area, approximately 1000 km in length and 200–300 km in width, covers an area of approximately 240 000 square km. It is characterized by shallow water (average 35 m), of high temperature with little fresh water inflow and its high evaporation rate results in highly saline conditions. The Gulf is connected to the Indian Ocean via the roughly 60-km-long Strait of Hormuz, which is only 20 km wide and 100 m deep. Eight countries border this sea area. They are Bahrain, Iran, Iraq, Kuwait, Oman, Qatar, Saudi Arabia and United Arab Emirates (UAE).

OIL POLLUTION SOURCES OF THE REGION

This region is the marine area that receives the largest quantities of oil pollution in the world. It is estimated that a total of approximately 144 000 metric tonnes of oil pollute these waters every year. Another estimate is that during the 10-year period from 1979 to 1989, about 150 000 metric tonnes of oil polluted the region annually. According to these estimates, oil pollution in the ROPME sea area represented 3.1 per cent of total world oil pollution, 47 times the average amount for a marine environment of a comparable size.

The major cause of oil pollution in the region is the discharge of ballast water. In addition the largest source of spilled oil in this regional sea comes from tanker transport accidents (57 per cent). The quantities lost in such circumstances are not precisely known. However, according to statistics compiled by the ROPME Emergency Centre, 13 confirmed incidents were reported between May 1981 and June 1987. They have caused pollution; 17 incidents have been classified as possible potential threats. By contrast, 82 *general* shipping incidents were recorded during the same period, their pollution having been confirmed in only 11 incidents.

A complete network of thousands of kilometres of pipelines lies in this sea. Those pipelines carry oil, gas and petroleum products from the offshore oil wells to shore facilities and terminals. Frequently, the submerged pipelines are ruptured and leak oil or oil products into the water. However, few data are available to evaluate the quantities of pollution from such accidents, as they often are not reported.

During the last 11 years, which included the Iran–Iraq war and Iraq–Kuwait war, much oil pollution was recorded. A total of approximately 750 000 barrels of crude oil was discharged into the sea from leaking Nowruz wells between February and September 1983 and 250 000 barrels from burning wells was discharged into the sea up to the end of May 1984. The total amount of oil that has poured out into the sea from the damaged wells of the Nowruz field over 16 months is estimated at approximately one million barrels or 150 000 metric tonnes. This is 2.5 times the amount of oil that was discharged from offshore oil pollution in all the other seas of the world in 1978.

THE LARGEST OIL SPILL IN HISTORY

On 19 January 1991, during the Gulf hostilities, most of the region's oil pollution sources discharged more than 5 million barrels of crude oil into the north of the Gulf Sea area, and the counterclockwise current and prevailing northwesterly winds guided the oil spill south along the Saudi Arabian Coast and contaminated a stretch of 600 km of the coast, which is a tremendously important ecological and economic area. The effect of this oil spill has harmed the marine habitats to a disastrous extent.

CONCLUSION

Why is this body of water one of the most polluted in the world? The answers to this question are manifold, among them the following:

1. More than 30 per cent of the world's marine transport of oil crosses the waters of this region. Between 20 000 and 35 000 tankers cross the Straight of Hormuz every year.

2. More than 25 major oil terminals are located in the area and the complex network of pipelines that lies at the bottom of this sea carries oil, gas and related petroleum products. Moreover, more than 800 oil-producing wells are found on the seabed of the region.

3. The management, legal and administrative mechanisms of the environmental protection agencies in the countries of the region are weak.

4. The region has been ravaged by war for more than 11 years.

5. The local and regional legislation and environmental protection laws, as well as the monitoring systems, where they exist, are not strong enough to deal with the amounts of pollution indicated above.

6. Very few countries of the region have joined the International Convention regarding the Protection of the Marine Environment from oil pollution.

7. Environmental awareness is low among the top decision-makers in the area.

Table 1   Marine Pollution Alert Reports from May 1981 to June 1987 in the Gulf Region (MEMAC, 1987, Bahrain)

War related incidents	
Pollution confirmed	13
Information N/A but possible potential threat	107
No pollution reported	126
Information N/A, not considered pollution potential	83
TOTAL INCIDENTS	329
General shipping incidents (from 1 January 1984)	
Problems on board (fires, engine failure etc.)	28
Grounding/touching submerged objects	20
Collisions	6
Sinking/capsizing	13
Illegal discharge	3
Incidents due to heavy weather	6
Miscellaneous: rupture of loading hose	6
Pollution confirmed	11
TOTAL INCIDENTS	82
Miscellaneous incidents	
(including oil-slick sightings)	15
Pollution confirmed	20

Chapter Eighteen

A brief overview of forests

REUBEN OLEMBO
Office of the Environment Programme, UNEP

Let us focus on forests because their role is so ill-appreciated. Previous figures of 6 billion hectares of forests world-wide were reduced to 4 billion by the late 1980s and are now even lower. The most prominent reason for this decline is human use, primarily for agriculture and the commercial exploitation of these forests' timber. Nearly 35 per cent of Earth's original forests have been converted to other uses because of the expansion of human activity. Apart from this, climate changes on a global scale are going to threaten forests even further. Available models indicate that a two-degree change in temperature will have drastic effects on forest cover and shift their areas, composition and productivity.

In terms of statistics, the major breakthrough in defining the extent of forest cover and size of deforestation was started in the 1970s by FAO, and has progressed in recent years. Recent data indicate that deforestation was progressing at a speed of 11.2 million ha/year. The greatest portion of tropical forests is in Latin America, while Africa has the world's largest tropical woodlands. Both types are affected by deforestation. We are hoping to reverse the current devastating trend by the year 2005.

The present conditions of forests and woodlands are very serious. Only recently have some governments started to realize that apart from the pure timber economic potential of forests, there is a broad range of other forest products which are of use to local populations. The central concept of the Tropical Forest Action Plan was an attempt to revise traditional perceptions and views on forest management. Although these intentions were good, the application of these new ideas was not completely satisfactory.

In a forest, what are considered "minor" non-wood products are not minor at all. Research carried out in Malaysia revealed that in those forests, 1290 non-timber species were used by human beings, not including their medicinal products. The commercial value of these products is far from negligible: in 1984 the export value of rattoons from Indonesia alone amounted to about US $90 million. Considering 10 more species, the total export value in the same year may have reached about one billion dollars.

In India, net revenues from forest products amounted in 1982 to

US $336 million, of which 40 per cent were from non-wood products. These figures are based on currently marketable goods. If one considers that a substantial part of forest products does not enter into the market cycle at all, the economic value of forests becomes even higher than trade figures show.

Another critical issue is the forest as a reservoir of untapped resources. Let us distinguish resources from ecosystems: the former are harvestable, while biodiversity is not harvestable. Available information indicates the potential of forest products for the manufacture of medicines such as analgesics, antibiotics, laxatives, tranquillizers, diuretics and new drugs to combat illnesses such as cancer. In 1985 the value of medicinal products then derived from forests amounted to $20 billion—without even considering the potential of forests as reservoirs of materials that have not been processed and refined. The environmental management of forests has to take into account these new elements. Another non-negligible aspect is the support role of forests to wild animals that are used as food. Information from West Africa indicates that approximately one-third of the meat diet of local populations is derived from wild animals. All this shows that national accounting systems are, at best, incomplete. Current statistics do not reflect these elements, which are so essential to planners. Hence the need for natural resource accounting at the national level.

Although the services provided by forests are well known, the monitoring of interventions is not satisfactory. For instance, with regard to reforestation, it is amazing that there is so little information on how many trees survive once they have been planted. This is a serious impediment to planning, the more so because any new treaty on forests must include a global agreement on reforestation. The establishment of an inventory on reforestation is also required by the prospect of developing agroforestry through the identification of optimal combinations of species suitable for such new enterprises. Unfortunately, experiences tested through traditional approaches have not tended to be recognized at the national level. Agroforestry is becoming so important within the framework of forestry management that agroforestry research institutes will rise to the level at which they can take on a combined global mandate. In many countries we therefore expect to see a networking of agroforestry institutes supported by such funding mechanisms as the International Centre for Research on Agroforestry.

UNEP EFFORTS TO COMBAT DEFORESTATION

UNEP activities to combat deforestation, undertaken in conjunction with sister United Nations' bodies, as well as major intergovernmental and non-governmental organizations, fall basically into four categories.

The first is sustaining the multiple roles and functions of all types of forests and woodlands because a rational and holistic approach to the sustainable and environmentally sound development of forests has yet to be realized. A prime activity in this context is the collection, compilation, updating and distribution of information on land classification undertaken within the UNEP/FAO Forest Assessment begun in 1980. Another is the development and implementation of plans and programmes—including national, regional and subregional goals—under the aegis of the Tropical Forestry Action Plan (TFAP). With international and regional cooperation and coordination, we may begin to arrive at a true valuation of forests. Finally, at the grassroots level, UNEP is promoting the participation of local communities in forestry—establishing, developing and attempting to sustain an effective system of forest extension and public education. One salient pilot project of this type, among a number of others, is taking place in Cajamarca in the Northern Sierra of Peru. There, with the local Development Corporation and the University of Cajamarca, the area's farmers have been encouraged to establish nurseries, launch afforestation programmes, carry out soil conservation and management and improve irrigation, human and animal health and further their own community development. In short, the project embodies a holistic approach.

The second area of UNEP effort is enhancing the protection, sustainable management and conservation and the greening of degraded areas through forest rehabilitation, afforestation, reforestation and other means of restoration. This thrust is based on the assumption that the degradation and conversion of forests to other types of land use are environmentally unsustainable. Among the ongoing activities in this area are revegetation in appropriate mountain areas (such as Cajamarca) and limiting and aiming to halt destructive shifting cultivation. UNEP is also working jointly with the National Environment Research Council of the United Kingdom on carbon sequestration—forests as carbon sources and carbon sinks. Lastly, in conjunction with FAO, the International Tropical Timber Organization (ITTO) and UNESCO, UNEP is helping to provide support for the management, conservation and sustainable development of forests, looking *inter alia*, to the conclusion of a new International Tropical Timber Agreement.

Third, because the vast potential of forests, forest lands and woodlands as a major resource for development has not yet been realized, UNEP is promoting efficient utilization and assesment to recover the full valuation of the goods and services provided by forests, forest lands and woodlands. In this connection, UNEP, FAO and the World Bank, together with the University of Minnesota's Forestry for Sustainable Development Program, have been developing reference materials on the assessment of forestry project impacts, defining issues and developing strategies, making economic assessments of forestry project impacts and valuing the multiple outputs of forests. Also, in conjunction with ITTO, UNEP is developing criteria and guidelines for the management, conservation and sustainable development of tropical forests.

Lastly, given the neglect of systematic observation and assessments of forest resources, management, conservation and development, UNEP is working to establish and strengthen capacities for such observation and assessments, including the use of forests in commercial processes and trade. This activity is being carried out largely through the UNEP/FAO Global Forestry Assessment and will continue with future assessments of all forest types.

THE FOREST PRINCIPLES OF THE UNITED NATIONS
CONFERENCE ON ENVIRONMENT AND DEVELOPMENT

Although most of the Forest Principles adopted by the United Nations Conference on Environment and Development (UNCED) are by no means new, they contain two noteworthy elements. First, they recognize that progress is not necessarily synonymous with moving forwards, that indigenous people are a source of tremendous wisdom in forest management. It is therefore enormously important to weave together traditional agricultural methods and crop varieties with new technologies in agroforestry.

Second, the cumulative emphasis of the Principles is their placement of deforestation and forest management within a wide context of economic factors including agriculture, trade and commodity pricing. Principle 9a states:

> The efforts of developing countries to strengthen the management, conservation and sustainable development of their forest resources should be supported by the international community, taking into account the importance of redressing external indebtedness, particularly where aggravated by the net transfer of resources to developed countries as well as the problem of achieving at least the replacement value of forests through improved market access for forest products.

It is these new approaches that may awaken governments to the need to overhaul revenue systems that subsidize the unsustainable profits of the timber sector at the expense of local people and long-term concerns. If this alone is achieved, we shall—with or without the proposed convention on forests—have taken a major step towards preserving the myriad functions of forests in the seamless web of the global ecosystem.

DISCUSSION

Question: Is there a working definition of woodlands as compared to forests?
Answer: The current criterion centres on the density of tree formation, i.e. number of trees in a given area.

Question: How can we improve our knowledge about the medicinal uses of forests?
Answer: One of the first steps we are taking within the biodiversity convention is to establish the biodiversity value of each country. Such assessment would include biodiversity as a medicinal resource.

Chapter Nineteen

Ecosystem improvement in the Guerrero Mountain region: a pilot project

JORGE MARTINEZ OJEDA
Ecosystem Development Division, Secretariat for Urban Development and Ecology, Mexico

The Guerrero Mountain region, like many other areas in Mexico, suffers from a shortage of food, basic social services (among them health services); and because of the adverse conditions of its topography has a deficient road infrastructure. Many of these problems are compounded by the variety of existing dialects. The background data for the project is given below.

Background of the project

Location:	The Mother Mountain Range South, bordered by the Pacific Ocean, between the states of Oaxaca and Puebla
Total area:	860 000 ha
Altitude:	Between 250 and 3000 m above sea level
Political division:	16 town councils
Population:	250 000 inhabitants (1990 census)
Annual migration to date:	19 000 inhabitants (7.6 per cent)
Population growth:	By the year 2000, 313 000 inhabitants (2 per cent annually)
Ethnal linguistic groups:	Foor, Nahuatl, Amuzgo, Tlapaneco and Mixteco

Until the project began in 1982, the local populations had tended to use the natural resources of the area inadequately and unsustainably to satisfy their needs. No recuperation plan existed, particularly to satisfy such basic needs as timber for housing and furniture, fuelwood for domestic use, and craft manufacture for both local consumption and national and international markets.

THE PROJECT OVERVIEW

From 1982, with low budgetary allocations, the project aimed at upgrading the living conditions of the region's inhabitants through road construction and the provision of health services. There have also been efforts to improve local farming to increase the local food supply.

The results to date, insufficient though they may be, stem largely from the General Law for Ecological Equilibrium and Environmental Protection, adopted in 1988, by which the Mexican government transfers power to the states and town council authorities, to prevent and control damage to the environment and to enhance the local capacity for solving local problems. Based on this law, the Secretariat of Urban Development and Ecology promotes the establishment of an Inter-institutional Agreement with the participation of the federal, state and town council authorities as well as research and educational institutes and community organizations. This Agreement establishes obligations for the elaboration and development of specific projects for soil recuperation, improving basic environmental health standards and basic food production, all linked to educational facilities.

So far, conservation measures on slopes have been taken to control landslides, primary health service centres have been built and, "Lorena"-type stoves that use fuelwood in an efficient way have been widely adopted. In addition, denuded tracts of land have been afforested with seedlings produced in local nurseries, environmental education promoters providing the necessary training for natural resources management. Finally, the Inter-institutional Agreement establishes commitments to the participants in accordance with their institutional responsibilities; these activities are programmed and results are checked periodically in evaluation sessions.

Inter-institutional Agreement Guerrero Mountain Project, Mexico

Participants:

Federal government:

- Secretariat for Urban Development and Ecology
- Secretariat for Agriculture and Water Resources
- Secretariat for Programming and Budget

State government:

- Secretariat for Rural Development

Municipal government:

- 10 Town Councils

Institutes of Research and Higher Education:

- National Indigenous Native Institutes
- National Forest studies at the Rural Investigations Institute
- National Autonomous University of Mexico

Communities of:

- Alcozouca, Metlatunuc, Atlauajalcingo Del Monte, Tlapta de Comonfort, Atlixtac, Huamuxmman Alpoyeca, Tlalixtaquilla, Xalpatlahuac and Copanotuyac

In addition, to ensure community participation, we propose the organization model presented below.

Community Organization Structure and Functions

1. *Town Council Environmental Protection Committee functions*:

    (a) Expedition of local regulations and definition of issues;

    (b) Development of the Town Council Environmental Protection Programme;

    (c) Budget proposals;

    (d) Inter-institutional co-ordination and convening of assemblies to this end.

2. *Ecological Protection Directorate functions*:

    (a) Development of Town Council Agreements;

    (b) Monitoring the regulations and norms relating to environmental pollution;

    (c) Promotion of the creation of a popular system for reporting infringements

3. *Ecological Council and Community Organization functions*:

    (a) Taking the role as an active entity to prevent environmental degradation;

    (b) Supporting the town council in the development of ecological programmes;

(c)   Increasing existing organized groups and adding new ones to conserve the environment.

Development planners in other countries may find these models useful.

Chapter Twenty

Forest resources in Thailand

AZIZ SAMOH
Southern Region Environmental Office, Office of the National Environment Board, Thailand

BACKGROUND

Rapid deforestation in Thailand over the past three decades has fuelled public alarm and concern over diminishing resources. Thirteen million hectares of natural forest have been denuded in 30 years, at an average rate of 0.43 million ha per year. Less than 28 per cent of the country (about 15.66 million ha) is now under forest cover. To protect the existing forest, government measures such as logging bans and stricter enforcement laws have been adopted to ensure that the forests will not be further encroached upon. These measures have been enacted on the assumption that the logging industry is the cause of deforestation.

Although it is undeniable that large-sized trees were initially cut down by loggers, villagers have followed loggers' trails and cleared medium- and small-sized trees to make room for farmland. From 1960 to 1990, Thailand's agricultural population increased by 14 million. During the same period, 13 million ha of forest were cleared. The problem of forest loss is no longer simply a problem of excessive logging: it is largely a problem of low-income villagers searching for agricultural land. The latest figures indicate that more than eight million people now reside in national forest reserves. These low-income villagers are legally termed "forest encroachers", although many had moved into the forests before they were declared forest reserves.

The environmental effects of deforestation are increasing in severity: flash floods in the south and droughts in the northeast are well known examples. Deforestation can no longer be tolerated, yet population growth continues and more farmland is needed. What needs to be done?

DESIRABLE ACTIVITIES AND OUTCOMES

1. An increase in forest cover to 40 per cent of Thailand's surface area, with an increased emphasis on community involvement in forestry programmes;

2. The division of Thailand's forests into 2 categories: protected forest (goal, 15 per cent of the country) and economic or productive forest (goal, 25 per cent);

Table 1 Natural forest areas of Thailand (Km²)

Region	Total land area	Area covered by forests				
		1961	1973	1982	1985	1988
North	169 644.3	116 275 (68.54)	113 595 (66.96)	87 756 (51.73)	84 414 (49.59)	80 402.3 (47.39)
East	36 502.5	21 163 (57.98)	14 876 (41.19)	8 000 (21.92)	7 990 (21.89)	7 833.8 (21.46)
North East	168 854.3	70 904 (41.99)	50 671 (30.01)	25 886 (15.33)	24 224 (14.35)	22 893.3 (14.03)
Central and West	67 398.7	35 660.5 (52.91)	23 970 (35.56)	18 516 (24.47)	17 228 (23.56)	17 244.4 (23.59)
South	70 715.2	29 626 (41.89)	18 435 (25.07)	16 442 (23.35)	15 485 (21.9)	14 629.6 (20.69)
TOTAL	513 115	273 628.5 (53.33)	221 547 (43.18)	156 600 (30.52)	149 341 (29.11)	143 003.4 (27.87)

Notes: All data were derived from LANDSAT, except for the 1961 data which came from aerial survey. Values given in parenthesis are percentages of the total land area in each region.

3. A revision of the laws and regulations governing the management of forest lands, especially in relation to private sector initiatives;

4. The development of coherent short, medium and long-term plans for forest and forest industry development;

5. The reform of forest administration to link it with these plans;

6. The identification and introduction of technical innovations designed to increase the productivity of forestry operation; and

7. The development of a public awareness programme to educate and inform people about the importance of forest resources.

FOREST PROTECTION PROGRAMME

To reach the target of the National Plan, the following strategies have been initiated:

Watershed classification and management

Since the land in the country's watershed areas varies from one place to another, watershed classification (or land-use planning for watershed areas) is needed. Watershed classification is an effort to make human uses of land as compatible as possible with the features of the environment and to mitigate adverse effects. Watershed classification will help to achieve this goal by identifying which areas should be maintained as protected forests and by prescribing guidelines for associated areas that may be used for the cultivation of trees or crops. Moreover, the watershed classification, as approved by the Cabinet, requires a system for establishing potential uses of land based on the physical characteristics of landscape units, namely elevation, slope, landform, geology and soil.

Watershed Class 1: protected forest (conservation forest)

Class 1A includes areas of protected forest and headwater source areas, usually as higher elevations with very steep slopes. These areas still remain under permanent forest cover. Class 1B are areas having physical features and environments similar to Watershed Class 1A, but where portions of the area have already been cleared for agricultural use or occupied by villages. These require special soil conservation protection measures and, where possible, should be reforested.

Watershed Class 2: commercial forest

Class 2 comprises areas of protected and/or commercial forest (usually the latter). For the most part, these areas are located at higher elevations, with steep to very steep slopes. Landforms are less erosive than in Watershed Class 1. Areas may be used for grazing or for certain crops. Soil protection measures are required.

Watershed Class 3: fruit tree plantations

Class 3 covers upland area with steep slopes and less erosive landforms. These areas are usually used for fruit tree plantations or certain agricultural crops and may be used for commercial forests, grazing or other purposes. They too require soil conservation measures.

Watershed Class 4: upland farming

Class 4 includes areas of gently sloping lands, suitable for row crops, fruit trees, and grazing, with a moderate need for soil-conservation measures.

Watershed Class 5: lowland farming

Class 5 groups gently sloping to flat areas, used for paddy fields or other agricultural uses, with few restrictions.

Multipurpose tree programmes and community forestry

Due to developing wood shortages over the last decade, fast-growing tree plantations have been promoted. These comprise about 26 tree species, among them pine, persian lilac, acacia, eucalyptus, leuceana, casuarina, duabanga, and acrocarpus.

The STK programme

One of the most important programmes is the granting of STK or "right to farm" land-use permits to squatters in national reserve forest areas. In principle, the STK programme is aimed at halting further encroachment upon forest reserves located near squatter settlements. The granting of STK gives the farmers concerned a sense of ownership they might not otherwise have had as well as an incentive to settle on and invest in the land they occupy. The STK programme also engages these individuals for the government to replant forests and maintain standing forest in the STK areas designated for agriculture.

Chapter Twenty-One

Kenya's policy on environment and development: focus on afforestation

BERNARD O. K'OMUDHO
Kenya National Environment Secretariat

Kenya is referred to as a land of contrasts. Whether this is stated from the standpoint of a tourist or a geographer, the country is endowed with varied physiographic features, a diversity of landscapes that support a wide variety of wildlife and other resources. Geographically situated in East Africa and on the Equator, the landscape varies from the snow-capped Mount Kenya at 5000 m, sloping eastwards to the shores of the Indian Ocean and westwards across the Great Rift Valley to the shores of Lake Victoria. Of the total land area of Kenya, about 80 per cent is classified as rangeland, which is arid to semi-arid, and where rainfed agriculture is limited by rainfall availability. Agriculturally, high-potential land is therefore small, though agriculture forms the backbone of the country's economy.

Since Kenya became independent in 1963, the pace of the country's development has been fairly rapid. This progress unfortunately, coupled with a high rate of population growth, tends to exert pressure on the available resources. Within the broad context of national development, environmental considerations have with time entered the mainstream of government policy, which aims at ensuring that available resources are utilized rationally. In an endeavour to achieve sustainable development, the government has been advocating proper environmental management or "development without destruction of the resource base". This is a long-term objective aimed at a life of harmony between human beings and their natural environment.

As early as 1965, the government spelled out measures to safeguard the country's natural resources, while pursuing development programmes to achieve greater heights of prosperity. The setting aside of protected wildlife areas and forest reserves was a milestone in the government's commitment to the cause of protection of natural resources. In subsequent years, the government laid emphasis on proper environmental management through the establishment of various institutions and programmes such as the foundation of the National Commission on Soil Conservation and Afforestation.

Policy on environmental protection, however, has tended to develop sectorally, based on the mandates of various government ministries and departments. The last development plan, 1989–1993, which emphasized the efficient utilization of resources on a sustainable basis led to the development of a set of comprehensive guidelines and strategies for harmonizing the formerly sector-specific policies into a coherent, holistic approach to protecting and managing the environment. This policy initiative set out broad objectives and strategies for achieving sustainable development. For instance, in conserving and managing natural resources, the goal was to attain and maintain environmental quality that permits a life of dignity and well-being for all Kenyans. And in promoting the maintenance of ecosystems and ecological processes essential for the functioning of the biosphere, the policy aimed at sustainable utilization of resources for the benefit of the present generations, while ensuring their potential to meet the needs of future generations, as recommended by the World Commission on Environment and Development in 1987.

Because public awareness and appreciation of the essential linkages between environment and development are important in achieving successful implementation of programmes, the policy advocated individual and community participation in all aspects of environmental management practices.

One of the main thrusts in implementation of this environment policy was the realization that vegetative cover has various accruing benefits, such as binding and stabilizing the soil, protecting essential watersheds, providing badly needed forest products like fuelwood and slowing climate change by providing a carbon sink, among many other functions. Moreover, fuelwood provides about 75 per cent of the energy needs of the country and accounts for about 95 per cent of the energy supply in the rural areas. The government therefore embarked on a nation-wide campaign of afforestation involving community participation, as well as increasing the acreage under plantation forests. More specifically, the major objectives of forest management are as follows:

- The protection of forests to maintain climatic and physical conditions of the country, to conserve and regulate water supplies and to conserve the soil by preventing desertification and soil movement caused by water and wind;
- The provision of fuelwood, charcoal, timber and other forest products, both for consumption within the country and for export;
- The provision of recreational facilities for the public and the preservation of wildlife; and
- The provision of employment.

The fundamental contribution of the forest areas is the essential role they play in protecting and enhancing the surface and ground-water supply in controlling soil erosion. Land under forest is made up of indigenous forest, forest plantations and rural afforestation schemes. The indigenous forests have been exploited for years and the policy now is that they should be protected from unwarranted destruction.

A systematic reafforestation programme of fast-growing exotic softwood plantations has been going on for many years, mainly to supply timber for domestic forest product requirements. Kenyans have now begun to see that the removal of forest cover leads to adverse environmental consequences. It reduces the water storage capacity of the soil, its infiltration capacity and fertility, and increases soil erodability, instability, stream flow and stream turbidity. The removal of forest cover can also adversely affect annual precipitation. With the decline in annual precipitation in some areas, evidence is emerging of a gradual change in the dominant vegetation types. Low forest, bush and thicket types of vegetation are giving way to coarse stunted grasses that are characteristic of arid zones. In some areas, the land is left bare, with little or no vegetation to protect the soil from the sun, wind and torrential downpours, leading to semi-desert types of landscape.

Recognizing these phenomena, many Kenyans have taken to tree planting as a measure to prevent environmental degradation, in particular soil erosion. Indeed, tree-planting has become so common an environmental activity that almost every public function is climaxed by ceremonial tree planting. Kenya is one of the African countries where the political goodwill and support for tree planting is well known. It is this factor that has acted as a catalyst in encouraging the people to plant trees.

Although tree planting and measures for the conservation of Kenya's forests are the responsibility of all the country's citizens, the Ministry of Environment and Natural Resources as well as some NGOs play a key role in promoting these activities. This is done mainly through the Forest Department, whose major activities include conservation, industrial plantation development and rural afforestation. In addition to the Forest Department, other government departments are actively involved in tree planting in rural areas. The Ministry of Agriculture runs tree nurseries in the districts, which raise tree seedlings for planting for soil conservation purposes. The Ministry of Energy operates a number of agroforestry centres. One of the roles of these centres is the raising of tree seedlings for planting in rural areas.

With about 80 per cent of the country falling in the arid and semi-arid zones, afforestation has by and large been faced with problems of tree seedling establishment in these arid areas, due to scanty and erratic rainfall and therefore low soil moisture. Developing the arid and semi-arid areas into economically productive lands utilizing their livestock, wildlife, water and the vegetation as essential resources remains a task for Kenya that must be carried out to ensure the sustainability of the development process. The recent creation of a new Ministry of Reclamation and Development of Arid and Semi-Arid Areas illustrates Kenya's commitment to making these areas productive lands in which resources are utilized rationally. The government has also established the Kenya Forestry Research Institute with a mandate, *inter alia*, of research on indigenous trees and suitable areas for their establishment, which is especially important for the country's arid and semi-arid areas.

There are several other organizations involved either directly or indirectly in the promotion of rural afforestation. These organizations include local non-governmental organizations, self-help groups, private companies and international donor agencies. The involvement of these organizations includes running tree nurseries, training local leaders and providing material assistance to groups that run the nurseries. We in Kenya realize that the destiny of our country will be determined by our own actions. What we do with our natural resources to ensure a better life and development is a serious thought for all Kenyans to reflect on as we look into the future. Many friendly countries and donor agencies have supported us in various programmes and projects to manage our environment and for this we are very grateful.

DISCUSSION

Question: Who owns Kenya's forests?

Answer: Most of the forests fall under government control, although a few belong to local authorities and several to private companies. Generally, the people think of the forests as government land and take little interest in planting trees or protecting them. Now, however, people are being motivated by being told that if they grow trees, they can earn money from so doing. In addition, forest programmes are being held, notably by non-governmental organizations, in the schools.

Chapter Twenty-Two

Village-level management of natural resources: a case study from the Himalayas

KESHAV DESIRAJU
Ministry of Environment and Forests, Government of India

INTRODUCTION

The Forest Panchayats (FPs) (from the Sanskrit *pancha* meaning "five") of the region known as Uttarakhand are good examples of village-level management of vital natural resources. Although they are highly region-specific examples, they teach more generally applicable lessons.

BACKGROUND

"Uttarakhand" is the name given to the Western Himalayas in the state of Uttar Pradesh. It consists of the sub-regions of Kumaon and Garhwal and has a total area of 53 123 sq. km. Roughly 37 per cent of this entirely hilly area is forest. The natural forests of the region are oak, pine and conifers. Several rivers, including the Ganges, have their sources in the snowy upper reaches of Uttarakhand.

The forest wealth of the region became subject to commercial exploitation with the growth of the railways in the 1850s. The Indian Forest Act of 1878 confirmed the government's monopoly over the forests with the demarcation of "reserved forests" (to date, the Forest Department controls 95 per cent of India's forests).

FORMATION AND FUNCTIONING OF FPs

These developments did not go unchallenged. As long as local residents had free access to forests and forest produce—and it was also a time when the population was very limited—the forests were not under any threat. The natural ecosystem could incorporate the requirements of the human (and cattle) population. With the reservation of forests and commercial felling, and the consequent loss of access, there were sporadic bursts of agitation and protest, leading ultimately to the Forest Panchayat Act in the 1920s. Forests considered to be of poor commercial value could henceforth be placed under the management of a locally elected body, with seven or nine members, and would be administered in accordance with rules framed by the government. The boundaries of the Forest Panchayat were clearly defined and maps drawn. The FP rules were elaborate and defined rights and powers, the nature of offences that could be dealt with by the village-level body and the punishments (usually fines) that could be imposed, the access to and the use of minor forest produce, etc. The control of the local administrative authority, who kept the account in which FP revenues were deposited and who also acted as a court of appeal, was absolute.

The members of the FP are required to meet periodically and maintain records and accounts. The principal source of revenue is on sale of resin from pine trees. Much smaller amounts can be made on the sale of grass. Most FPs also have their own system of sharing grass and other produce among all the villagers. These non-monetary transactions are conducted entirely at the village level.

Panchayats were originally created on degraded sites or poor forests. Sometimes a forest already in existence and protected by the villagers as "sacred" formed the core of the FP. A recent estimate states that 4058 FPs exist in Uttarakhand, covering an area of 220 000 ha. Some very successful FPs are known, where rich mixed forests exist, often much better than neighbouring reserved forests. Others are not so successful and are subject to encroachment and grazing. Still others are rich because of resin but are exclusively pine forests.

LESSONS:

Several important lessons can be learnt from the experiences of the Forest Panchayats.

1. Village-level management of natural resources is clearly possible. The institution of the FP was by no means ancient. It was created by government order. However, it obviously came close to some easily understood ideal. Further, it is not necessary that an NGO or some other outside group provide the catalytic effect. Within the community, the necessary leadership and management skills exist and are waiting to be used.

2. Financial management skills, in particular, can also be noted. Most FPs have accounts painstakingly maintained even, or especially, where the amounts involved are very small.

3. A strong sense of commitment to the Forest Panchayat exists. Villagers will usually protect the boundaries of their forest and keep outsiders (and their cattle) away. This contrasts sharply with the attitude towards "reserved" forests, which are perceived as alien. A large number of FPs employ guards, who are paid from their revenues. The fixed wage is usually very small, but the guard has recognized duties.

4. Women have been more articulate and forceful in their involvement with the

FP than in other village-level institutions. While relatively few women are actually elected members, they do have a say about what is to be planted and also are actively protecting the forests. Women in this region do *all* the collection of fuelwood, fodder and water and *most* of the agricultural work. Despite this, their role as producers in the economic cycle is largely unrecognized. In this context, their role in FPs' affairs is important.

5. Forest Panchayats have survived *only* where pasture land exists separately and where some sort of wall or other protection exists. Recent efforts to increase areas under plantation by including traditional grazing lands have been uniformly unsuccessful. There is a clear understanding of differing land uses, and the requirement of separating agricultural, forest and pasture land. This is reflected also in the willingness of most FPs to evict persons who have forcibly begun cultivation or also have built houses within the forest.

The usefulness of protection is also self-evident. Even the simplest kind of stone wall can have a noticeable impact on plant survival rates, and the growth of fodder grasses. Having said this, one must recognize the problem issues that remain:

1. The source of FP revenues and the management of these funds are both fraught with difficulties. Resin from the Himalayan pine is a major income earner. Pine grows easily, regenerates naturally and can also be planted. It is not eaten by grazing animals. However, the growth of pine forests rules out the growth of any sort of undergrowth or of any other plantation. Villagers are well aware of the advantages of mixed forests, but as income yields from such forests can be expected only well in the long-term future, the temptation to plant pine, or to allow its natural growth, is great.

   One solution could be a ready and plentiful supply of saplings, along with a sustained awareness campaign. The supply of saplings is the responsibility of the Forest Department, though some FPs actually have their own nurseries. Another solution could be to encourage more women to become active members of the FP. As indicated earlier, the physical burden of collecting fuelwood and fodder rests on women of the village and where they have had a say, such species have been planted in the Forest Panchayats, leading to a saving in their time and energy. The benefits of cash income, as from resin, for example, are seen only by the men.

2. The question of management is also delicate. The existing rules place all the resin revenues in the control of the local administrative authority, leaving the FP only with petty amounts raised from fines, the sale of fodder grass and other minor forest products. How the local authority exercises its authority, whether funds are released to the FPs on demand, in time, and for what purposes all become major problems. The answer obviously lies in delegating all these powers, but hitherto, the government has not thought along these lines at all. There has been no indication that government would like to relinquish control.

3. Another problem area is the funding of poor Panchayats. Given that the organizational structure and skills exist, Panchayats that do not have substantial resin revenues are usually unable to do very much beyond protecting their land. Well-motivated government schemes have sought to make funds available, but this immediately raises questions of accountability, the use of public funds, the need for audit, etc. It is certainly not the case that government funds given to FPs would inevitably be misused. However, our present administrative arrangements do not readily allow such transfers. This is being done in FPs and under other major schemes to fund NGOs in the area of wasteland development, but the motivations have yet to be internalized in government procedure.

The role of the government in village-level activities itself becomes a question that administrators and managers have to ask themselves. Where the need is felt and perceived, where the ability and the skills exist, the government's role should ideally be that of a facilitator. The responsibility for decision-making, or even for making mistakes, should be given to the persons actively working. This principle applies as much to welfare or rural development schemes, as it does to the community forestry. There is also the disturbing feature that at the village level, the government is not often seen as a benefactor but as a source of harassment and corruption. This is an image that must change.

Lastly, there is a great urgency. It has been recognized that in a country of India's size, the scarcity of biomass reserves constitutes *the* major issue. As two of India's more enlighted activists have observed ". . . the key objective of rural development programmes must be to restore ecological balance and increase biomass production on a sustainable and equitable basis" (Anil Agarwal and Sunita Narain: *Towards Green Villages*, 1990). The FPs of Uttarakhand are an example of an institution that can take on the challenge, if allowed to function as a representative body with the necessary powers. Obviously this pattern may not work elsewhere. Every region has its own ecosystem (India can be divided into 16 major ecosystems) and patterns of land availability, land use, tenure and cultural practices vary enormously. However, if the approach is to think in terms of villages, and of people living there, then every village may find its own solution.

Chapter Twenty-Three

Land degradation and desertification in Ghana

ABDULLAH IDDRISU
Environmental Protection Council, Ghana

Ghana is a tropical country lying between latitudes 4° 45'N and 11° 15'N and longitudes 3° 15'E and 1° 15'E. Its total area is 238 533 sq. km, representing 0.8 per cent of the area of the African continent. The last census in 1984 reports Ghana's population at 12.2 million. Official estimates, however, put the population at 14.4 million in 1988 and 15 million in 1990, with a crude current density of 62.8 persons per sq. km.

Ghana has a broad natural resource base, whose utilization has been both beneficial and adverse. Agriculture is by far the dominant economic activity, accounting for about 51 per cent of GDP and contributing over 50 per cent of the country's total export revenues. It also provides employment for over 60 per cent of the working population. Ghana is classified as a low-income Sub-Saharan African country with a per capita GNP of US $370 (1987) and an annual economic growth rate of 1.4 per cent (1987, at purchaser values).

CAUSES OF DESERTIFICATION

The area of Ghana that shows marked signs of desertification is the Savanna Zone, which occupies two-thirds of the country. The areas of Ghana threatened by desertification are:

- the area of the country north of latitude 7°30' and east of longitude 1° 45';
- the northwestern corner of Ghana;
- the Coastal Savannah of the Central and Greater Accra regions.

The causes of land degradation that lead to desertification in Ghana can be grouped into human and non-human factors. The non-human factors are long-term climatic changes and cyclic fluctuations. This has become evident in unreliable and variable rainfall that, *inter alia*, affects rivers, which dry up during certain drought years. The human factor manifests itself in the form of destructive cropping practices that encourage soil erosion and soil degradation; the removal of vegetation for fuel, construction, medicine and other uses; the improper logging of timber for export; the burning of vegetation for land-clearing and hunting (bush fires); and population and livestock pressure on the land beyond its carrying capacity and overgrazing by livestock.

The stocking rate of animals, especially cattle, in some parts of the country is estimated to be 1.3 ha/cow, which is far in excess of the estimated carrying capacity of about 10–12 ha/cow. These areas consequently are overgrazed, which results in the creation of bare patches that are subject to accelerated water and wind erosion. Transhumance is another cause of land degradation leading to desertification.

EFFORTS TO COMBAT DESERTIFICATION IN GHANA

Once it sets in, desertification has negative impacts. Periodic droughts in the savanna areas in Ghana have impoverished the people immediately affected. Reduction in meat supply is one result of desertification, due to the death of livestock caused by inadequate fodder and low-quality feed due to overgrazing. The wide spread of infant diseases such as measles due to general undernourishment and diseases such as kwashiorkor, directly due to protein–calorie deficiencies, are other alarming impacts.

Recognizing the negative impacts of desertification on the environment and the general well-being of Ghanaians, the government of Ghana has initiated actions aimed at combating desertification. Some of its efforts at combating drought and desertification are the following:

- In 1984, Ghana's Environmental Protection Council (EPC) set up a Committee to plan a National Workshop on Drought and Desertification for the country. Held in Accra in 1985, the Workshop reviewed actions which had been taken in Ghana to combat drought and desertification and drew up plans and strategies for action to combat desertification. In the same year, the EPC transformed the Workshop into a National Planning Committee to prepare a National Plan of Action to combat the effects of drought and desertification (NPACD).
- In 1987, the Plan of Action was completed, identifying priority anti-desertification projects for which Ghana wanted the assistance of the United Nations' Sudano–Sahelian Office (UNSO) to mobilize resources for action.
- A Desertification Control Unit has been set up in the Environmental Protection Council with a field office in Bolgatanga and an UNSO desk has been created at the United Nations' Development Office in Accra for desertification work in Ghana. The Unit is expected to undertake monitoring and evaluation of anti-desertification activities and programmes in the country and also to provide a mechanism for the coordination of the activities of agencies implementing programmes for combating desertification. Education and dissemination of information on the causes and problems of drought and desertification are major objectives of the Unit.

- A National Environmental Action Plan for Ghana has been prepared. It specifies that relevant government agencies and institutions dealing with issues concerning drought and desertification are to be strengthened to enhance their capability and capacities in implementing the anti-desertification projects identified in the National Plan of Action to Combat Desertification.
- An Environmental Awareness programme has been launched in the target areas and among target groups, such as farmers and pastoralists, on rational land management.
- The government encourages the establishment of woodlots and community wood plantations as sources of energy supply for the local people.
- An electrification programme has also been embarked upon by the government, with the aim of making hydro-electric power available to all major towns and villages in Ghana. This programme, when completed, will help reduce the burden on the country's forests for the supply of wood for domestic energy.
- In addition, the government has embarked on other energy conservation efforts such as the introduction and promotion of energy-saving devices, especially for cooking. Improved charcoal stoves with charcoal-saving efficiency of 42 per cent over the traditional coal port have been designed and produced in Ghana under the auspices of the Ministry of Energy.
- An improved charcoal-making technique known as the "Cassamance Kiln" has been designed, tested and is being promoted, since the technique helps in achieving efficiency in fuelwood in the charcoal production process.

These are but a few of the many efforts being made to give the support necessary to projects that are anti-desertification-oriented throughout Ghana.

DISCUSSION

Question: When did desertification begin taking place in Ghana?
Answer: The government recognized the northern part of the country as a desert in 1974.

Question: Is the Forestry Department involved in combating the problem?
Answer: The Forestry Department is playing an active role by providing free facilities and other incentives for reforestation of the area, as well as regularly monitoring the reforestation programme. One community organization has begun a significant reforestation programme. However, once fuelwood is exhausted, most poor communities simply leave the area. They do not regard the problem as theirs and hold the government responsible for all actions. None the less, efforts are being made to launch an ownership programme so that such people can be motivated to contribute to the reforestation programme.

Chapter Twenty-Four

The problem of desertification in Pakistan

MAHTAB AKBAR RASHDI
Environmental Protection Agency, Government of Sindh, Pakistan

The process of desertification in Pakistan falls into three categories:

- the expansion of the Great desert of Thar and Cholistan adjacent to the Indian desert of Rajasthan;
- the expansion of desertification due to deforestation or the felling of trees without any justification;
- land lost because of waterlogging and salinity.

The desert of Thar and Cholistan, which is expanding rapidly, was estimated at 120 000 sq. km in 1980. Its expansion has not been measured, as the environmental protection agencies in the country are new and have not yet acquired the technical expertise for this task.

The desert is rich in wildlife; in 1956, Pakistani scientists, with the help of the Gazetteer of the British government, identified 3000 species in the Sindh Area of this desert. At present (1991), species loss is extreme, as there was no rain during the whole year. This may be due to climate change intensified by the mushrooming of industries in the area which have increased the carbon-dioxide level of the atmosphere.

Added to the absence of the usual torrential rains is the problem of fresh-water resources. The ground water level in some of the oases of the desert is 60 to 100 metres deep; consequently, it is not possible to supply enough drinking water for human consumption.

When the rainfall began to decrease in 1987, the government of Sindh developed a project for constructing small dams to store water for irrigation and human consumption. Although the local component for the project was available, the donor agencies not only refused to finance this plan but also failed to provide any alternative suggestion to combat the process of desertification.

Another major cause of the expansion of desert is the removal of vegetation and trees. The reasons for this range from the demands of urbanization in the absence of any other cheap fuel to the need to eliminate bandit hideouts. This process began some 50 years ago when the Sanghar forest was removed by the colonial authorities to crush the insurgency of freedom fighters. Unfortunately, this process has continued as the forest continues to provide cover to undesirables. Further, urbanization has affected both wetlands and the forest close to the developing settlements. Lastly, the wood forests provide cheap fuel for the rural population. In areas where the rural population has an annual income of US $200 or less, no government or environmental body can motivate the people not to use fuel wood until and unless resources are made available to purchase cheap liquid fuels or cylinder gas.

The canal irrigation system begun in the central and southern areas of Pakistan towards the end of the 1930s has resulted in waterlogging and salinity in the area around the canals. About 20 per cent of the irrigated land was lost by 1960. Efforts have been made by the government to control the loss of land and to reclaim land through a programme of on-farm water management and salinity drains. Because of extensive efforts, whose costs were partly shared by the developed countries, the loss of land has been slowed down to 1.5 per cent annually. However, much more aid and technical assistance is required to reclaim the lost land.

In its resolution 44/172, the General Assembly directed UNEP to call in the technical experts and combat desertification. The implementation of this resolution, however, has remained largely limited to Africa. Therefore, the Economic and Social Council adopted resolution 1991/97, which asked UNEP to intensify efforts to combat desertification in the Asian region.

No groundwork has commenced as yet, despite human suffering, wildlife extinction in the area and other environmental stress. Action is long overdue. If and when life no longer exists in an area, let it not be said that the government of Sindh did not try to halt desertification and its consequences.

Chapter Twenty-Five

The National Conservation Strategy (NCS) for Pakistan

MOHAMMAD ASLAM MALIK
Environment and Urban Affairs Division, Government of Pakistan

Pakistan suffers from the plethora of environmental problems that generally plague newly industrializing countries, ranging from the depletion of renewable and non-renewable resources to a lack of enforcement of existing laws related to the environment. Moreover, although the government's Environment and Urban Affairs Division has the status of a ministry, the subject of environment *per se* is not included either in the nation's Seventh Five-Year Plan (1988–1992) or its Perspective Plan for the Year 2000. Consequently, it is extremely difficult to formulate concrete programmes. All environmental activities become budgetary "extras".

To overcome these problems, to ensure the sustainable use of natural resources, to preserve the environment and genetic diversity and to maintain essential ecological balances in the country, the government recognized the need to have a comprehensive National Conservation Strategy (NCS) in the early 1980s. The International Union for Conservation of Nature and Natural Resources (IUCN) fielded a mission to Pakistan in 1983 and, after detailed discussions with the government authorities concerned, agreed to undertake the NCS exercise for Pakistan. The Canadian International Development Agency (CIDA) provided funding for this study.

Accordingly in 1988, the government entered into an agreement with IUCN to develop the Pakistan NCS, which would identify how economic growth and several developments would be promoted, in the long run, by protection of environmental quality and rehabilitation of environmental damage.

The NCS Secretariat was established in the Environment and Urban Affairs Division under the supervision of a high-level Steering Committee, headed by the deputy chairman of the Planning Commission with eight federal Secretaries concerned with natural resources and five distinguished citizens as its members. This Steering Committee continuously monitored the development process of the NCS through July 1991, when the NCS document was finalized.

MAIN OBJECTIVES OF THE NCS

To be successful, large and complex endeavours require explicit objectives. The NCS has three:

- maintaining the essential ecological processes and life support systems on which human survival and development depend;
- preserving genetic diversity, on which many of the above processes and life support systems depend: the breeding programmes necessary for the protection and improvement of the cultivated plants, domesticated animals and micro-organisms, as well as advancement in scientific and technical innovations and the security of industries that use the living resources;
- ensuring the sustainable utilization of species and ecosystems, notably fish and other wildlife, forests and grazing lands, that support millions of rural communities, as well as major industries in the country.

We can summarize these objectives under existing goals, notably:

- conservation of natural resources;
- sustainable development; and
- improved efficiency in the use and management of resources.

Main operating principles

The three main operating principles of NCS are the following:

- achieving greater public partnership in development and management;
- merging environment and economics in decision-making;
- focusing on durable improvements in the quality of life.

NCS DOCUMENTS

The NCS document is divided into three parts and 13 chapters:

I. *Pakistan and the environment*
   1. Pakistan and Global Environment Changes
   2. The State of Pakistan's Environment
   3. Resource Use Impacts and Linkages
   4. Existing Institutions
   5. Present Policies/Programmes for Environment

II. *Elements of the National Conservation Strategy*
   6. Objectives, Principles and Instruments

KEY PROGRAMMES FOR IMPLEMENTATION

The NCS recommends 14 key programme areas/sectors for priority implementation:

1. Maintaining soils in croplands;

2. Increasing irrigation efficiency;

3. Protecting watersheds;

4. Supporting forestry and plantation plans;

5. Restoring rangelands and improving livestocks;

6. Protecting water bodies and sustaining fisheries;

7. Conserving biodiversity;

8. Increasing energy efficiency;

9. Maintaining and developing renewable resources;

10. Preventing/abating pollution;

11. Managing urban wastes;

12. Supporting institutions for common resources;

13. Integrating population and environment programmes;

14. Preserving Pakistan's cultural heritage.

Sixty-eight specific programmes have so far been identified in the above areas, each with a long-term goal and with expected outputs and physical investments required during the next decade (1992–2001). Each programme also has communication, extension, research and training components.

IMPLEMENTING STRATEGY

The NCS is a call for action addressed to senior and local government officers, businesses, NGOs, communities and individuals. The main points of the implementation strategy are the following:

- Revision of existing national plans in various sectors;
- Development of new plans;
- Incorporating NCS policies and measures into the Five-Year Plan (1993–1997);
- Incorporating NCS recommendations into various sectoral plans for agriculture, irrigation, and watersheds; into the forestry master plan; and into industry, science and technology and population policies;
- Allocation and reappropriation of funds;
- Development and strengthening of institutions;
- Development of environmental awareness and political will at all levels.

Impelementing agencies levels

- Federal and provincial leadership (ministries and departments)
- Departmental responsibility;
- District coordination;
- Community participation;
- Individual actions;
- Corporate tasks, and
- Government/NGO support.

Funding for NCS implementation

The NCS itself is not an investment plan. Instead, it is a guideline concerned with the policies for investments in various sectors. The recommended 14 programmes involve an overall additional investment of more than US $6 billion over a period of 10 years, i.e. 1992–2001. However, it does not require expenditure of *more* money, but, rather, calls for channelling investments towards maintaining and enhancing natural resources and towards increasing efficiency in the use of critical renewable and non-renewable resources. If implemented efficiently, the NCS will contribute towards financial self-reliance and sustainability by maintaining the capital stock of resources, allowing the withdrawal of short-term starter funds.

THE PRESENT STATUS OF THE NCS

Currently, an NCS Coordination Cell has been created in the Environment and Urban Affairs Division. An Executive Summary of this 620-page document has been prepared, together with another Summary for the approval of NCS by the cabinet after comments from the provincial governments have been received. The implementation phase started in 1992 for a period of 10 years to achieve the objective of sound and sustainable development and protection and preservation of the country's environment on a long-term basis.

DISCUSSION

Question: How is the strategy being implemented without external aid?

Answer: Currently, 4 per cent of national resources are being spent on the programme, while an additional 4 per cent have been set aside for the next 10 years. The government regards this as a type of savings, since the environmental resources will eventually pay back far more than the investment.

Question: Is Pakistan sharing its information and views with other countries?

Answer: Yes. The Sindh Arid Zone Development Authority (SAZDA) has been established with the cooperation of China. Pakistan also looks forward to assistance from and cooperation with other countries.

Chapter Twenty-Six

Rhino and elephant management in Zimbabwe

KESARI S. CHIGWEREWE
Department of National Parks and Wildlife Management, Ministry of Environment and Tourism, Zimbabwe

The expression "wildlife management" is largely a euphemism for the blunt fact of killing elephants. It does, however, imply that they are killed according to a plan, with objectives and technical data to support decisions.

Zimbabwe has been under more pressure than usual for managing its elephants. People ask, "Why, when the species is on the verge of extinction, can one country afford to carry out such apparently anti-conservation activities?" The subject is indeed complex. And it is difficult to give a simple answer that deals with the complexities as does the simple demand for an ivory trade ban that appears to satisfy much of the world's public. The Zimbabwe Department of National Parks and Wildlife Management is far from entrenched in its position on culling; it would be the first to avoid culling if it felt that culling were unnecessary or if it could be demonstrated elsewhere that the elephant population increase posed no threat to the environment.

CARRYING CAPACITY FOR ELEPHANTS

Many conservationists believe that wild animals possess all the necessary intrinsic mechanisms to limit their numbers at a level that is in balance with the environment. There should be no need for human intervention because the "balance of nature" does the job. Undoubtedly, large mammals are ultimately limited by their food resources. Food production, in turn, is limited by rainfall and soil fertility; this effect is striking in the semi-arid savannas of Africa. However, there is no recorded case in which an elephant population has increased in an orderly manner to an asymptotic level and remained there, comfortably regulating its birthrate in accordance with its death rate. Long after elephant numbers have exceeded the level at which food consumption matches vegetation production, populations continue to grow. At this stage they are eating into the "capital" of the system in a non-sustainable manner. When the crash comes, it is spectacular—as in the case of the

Tsavo National Park in the early 1970s, when a drought resulted in the death of thousands of animals. There are risks that other species may become extinct as elephants destroy their habitat, and the terrain cannot be recolonized by nearby populations because the given protected area is already an ecological island.

My own view of the relationships between elephants and vegetation is as follows:

First, it is important to recognize that elephants are the single most important factor that shapes the face of ecosystems where they are present. They are capable of converting woodlands into shrublands, shrublands into grasslands and grasslands into bare soil.

Second, in most Zimbabwean vegetation types, mature trees are likely to disappear completely when elephants live at densities greater than one per sq. km. Although this may not appear to constitute an ecological problem, it is difficult to see how certain birds and invertebrates that live only in tree canopies can continue to exist.

Thirdly, as densities continue to increase, a point may be reached where elephant food consumption exceeds the sustainable production of the system. Usually some episodic factor, such as a major drought, will result in a population crash of elephants. But this may not happen before irreversible changes have occurred—including the possibility of a permanent loss of topsoil through erosion.

OBJECTIVES FOR LAND WITH ELEPHANT POPULATIONS

Almost all Zimbabwean parks have the common objective of conserving a broad spectrum of large mammals and their habitats. The emphasis is generally on maintaining entire ecosystems in a positive state rather than favouring any single large mammal species at the expense of others. Inevitably, this results in a need to manage elephants.

METHODS OF MANAGEMENT

Zimbabwe manages its elephants not to any predetermined density but rather according to the desired result of habitat condition. This implies using the concept of adaptive management: elephants are reduced by an amount that simulation models predict will achieve the desired result and the effects are monitored by ecologists.

Elephant numbers can be reduced only by removing breeding herds. In a population that increases at 5 per cent annum, the recruitment to the adult male segment is less than the overall annual increment and the population would continue to increase regardless of how many adult males were killed—except in the extreme case where there were no males left for breeding. This is why sport hunting cannot substitute for the culling of breeding herds. Zimbabwe uses the technique of removing entire cow herds, on the assumption that this leaves the age structure of the general elephant population undistorted and causes the least trauma to the animals. One effect of removing cow herds is that adult males are excluded. Males have a

Table 1   Approximate Elephant Population of Selected Countries of Africa

Country	Approximate elephant population
Angola	20 000
Botswana	65 000
Malawi	2 000
Mozambique	20 000
Namibia	5 000
South Africa	10 000
Zambia	40 000
Zimbabwe*	65 000–75 000
* Carrying capacity, Park	35 000
Estate	10 000
Elsewhere	20 000

higher mortality rate than females, and males also tend to be killed outside protected areas as a result of crop raiding.

If one examines the example of a population of 10 000 elephants, the important features of its management emerge as the following:

- the population would increase naturally at a rate of 5 per cent per annum in the absence of management;
- an offtake of 4.1 per cent of the population in the form of breeding herds is sufficient to keep the population from increasing;
- an offtake of 0.2 per cent of the population consisting of adult males that are problem animals is killed outside the protected areas as crop raiders;
- of the adult males, 0.7 per cent are taken as sport hunting trophies. This is the maximum percentage that, after allowing for the problem of animal control, is compatible with producing large tusks for the hunting market.

RHINO MANAGEMENT

Zimbabwe's rhinoceros management is three-pronged: its first aim is to protect the wildlife population and consists of anti-poaching activities and undercover surveillance to prevent rhino trafficking. Second, the Department of National Parks and Wildlife Management translocates rhinos from border areas to safer areas. The minimum number in safe areas is 50 animals, to allow for breeding. Hence the Department holds conservancies at some 350 animals. Third, captive breeding programmes are carried out.

These are divided into two categories: designated *in situ* centres within Zimbabwe and the Chipangali Wildlife Orphanage and Boulton Atlantica Park. Zimbabwe's conservation activities are carried out in *ex situ*, in cooperation with international zoos. For example, the Black Rhino Foundation of the USA should have 40 rhinos at the end of our current joint programmes. This number should breed to 200 animals. These rhinos remain the property of Zimbabwe. If they breed successfully, the progeny will be relocated into the wild.

DISCUSSION

Comments: Several participants strongly queried culling as a means of management of elephant populations.

Question: With the scientific control of pruning the elephants by culling them, isn't Zimbabwe going against the international spirit of protecting endangered species?
Answer: We are trying to avoid ecological clash as a result of sustainable management.

Question: With the banning of trade in ivory and related products, will Zimbabwe stockpile these products without selling to external markets?
Answer: In the spirit of internationalism, we are seeking permission to suspend the agreement until the issue can be resolved.

Chapter Twenty-Seven

The built-up ecosystems

NAIGZY GEBREMEDHIN
*Chief, Technology and Environment Branch, UNEP**

The first question always asked when discussing this subject is this: is the built-up environment an ecosystem? Second, can we apply the principles of ecosystems to it? People usually see the built environment as an intrusion into the natural environment. Such a perception is not totally acceptable. UNEP is working to overcome this concept of antagonism between natural and built environments, with a view to harmonizing these two dimensions of the human environment. This conception is embodied in a programme entitled "The Environmental Aspects of Human Settlements", on which we are working with other United Nations' bodies, particularly the Centre for Human Settlements, *Habitat.*

One must constantly remember the fact that the built environment will keep growing, while the natural environment will either remain the same in scale or, more probably, diminish because of human pressure and activities. Urban population is increasing in percentage very rapidly *vis-à-vis* total population. The rate of growth of urban population typically ranges between 2.5 per cent the lowest rate (which is found in Jamaica) up to more than 7 per cent, which means that each 11 years, population doubles. Because of such expansion, human relations with the natural environment are very significant. The unplanned sector of urban growth is increasing even more rapidly than the planned sector, putting enormous pressure on public services and structures.

Spontaneous squatter settlements often result from the failure of the formal sector to provide proper shelter. Such settlements are often used as political tools and consequently become major political controversies. The formal sector sometimes reacts by using force to uproot spontaneous settlements and relocate their populations. A new approach stipulates the creation of a process in which the government and squatters work together, the government realizing its inability to do everything and therefore guiding the expansion of spontaneous settlements by developing a variety of policies, such as incorporating spontaneous settlements within site and service schemes, which decongest urban areas and rebuild city zones, and providing a reasonable level of basic services.

An unmanageable rate of population growth lies at the heart of the dilemma. The governments and societies of many developing countries cannot resolve this

---

* *Presently* Environmental Co-ordinator, Eritrea

crisis. It is therefore unrealistic to expect urban growth to slow down in the short term. Unfortunately, there are few successful solutions. Decentralization may help; policies aimed at fostering equitable development throughout the nation may help; a careful and balanced siting of economic activities may help. However, these are long, complex processes and, in the short run, we are likely to face a further aggravation of this problem.

In environmental management, density is a very important factor. In Cairo, gross density was 24 000 people per sq. km in 1976; in Manila it was 43 000 per sq. km in 1980. What such density means in terms of access to sanitation services has been vividly described.

The management of urban wastes is also very important. Typically, per capita waste produced in the urban centres of developing countries is very low. In Ghana, it is 0.8 kg per day, and in Kenya about 1 kg per day. With growing per capita income, the trend towards waste production is increasing.

The composition of wastes, however, also depends on development levels. In the poorer countries, most urban wastes are composed of vegetable products, while there are only small quantities of paper and plastic wastes. Such information is vital for the design of optimal waste disposal systems. For instance, in the case of dense wastes, which are typical of vegetable residues, it makes sense to buy compactor vehicles that reduce volumes and ease transportation and landfill space shortage. Such information is also critical to decisions concerning how to treat wastes. For instance, wastes with high concentrations of heavy metals, paper and textiles cannot be efficiently utilized in compost plants. Likewise, wastes with a high percentage of vegetable products, high moisture and little volatile matter are not suitable for incineration plants.

What kind of planning intervention approach is possible for the built ecosystems? At UNEP, we believe that it is possible to apply the principles of natural ecosystems' discipline and sustainability. The question of sustainability is of paramount importance. We must identify solutions which permit establishing sustainable equilibria between production and consumption, waste and recycling.

DISCUSSION

Question: Are there situations in which the growth of human settlements can contribute positively to the natural environment?
Answer: Yes. Cities are not inherently unproductive in environmental terms. Consequently, they can be planned to benefit the natural environment. One of UNEP's goals is reducing the types of subsidies cities now receive from the natural environment. One must also ask another question that reflects the new approach cited above. Who is responsible for creating a sustainable built environment? Governments, the people—or a combination of both? Herein lies a concept of enablement: people have a primary role in creating their environment, while governments should play a facilitating role that will allow people to realize such objectives.
Comment: Rural development and improvement can and do help reduce urban drift

and re-establish a balance between cities and the countryside so vital to the health of a nation. However, enhancing progress in rural areas is linked to a variety of issues, among them the question of debt payments. Resources are often allocated where progress is most visible. Political decision may therefore distort the true priorities, paying inadequate attention to the needs of rural areas.

Comment: In addition, centralization policies are an important contributory element. Reversing such policies is often difficult because of powerful interests that do not want to relinquish their power.

Comment: It is also important to note that management of urban centres often falls between central and local government responsibilities. Such situations also cause conflicts.

Chapter Twenty-Eight

Waste management in Uganda

PATRICK ISAGARA KAMANDA
Environment Officer, Department of Environment, Uganda

INTRODUCTION

In Uganda, a country that enjoys a climate favourable to human life, the main economic activities are crop production and livestock production. About 70 per cent of the country's income derives from these two activities. However, the country's good climate has some disadvantages as well, notably that of providing good breeding conditions for disease vectors such as mosquitoes and tsetse flies, which can retard crop production as well as reduce livestock, if not properly checked.

To protect the animals and to increase crop yields, thousands of tonnes of chemicals are imported into the country yearly, among them fertilizers, herbicides and pesticides. These chemicals eventually pass into the environment, either in their original composition or in a modified form, polluting air, land and water through application or disposal after use. It is not uncommon for untreated wastes from cattle dips to be disposed of directly on land and in water bodies.

Various municipal wastes also pose an increasingly significant problem as the population in urban areas continues to grow. In slums people cohabit with heaps of garbage. These individuals and families therefore live in poor health conditions, which raise the country's mortality rate and lower life expectancy, the latter now estimated at about 45 years.

The national income of Uganda has been dwindling because of the continuing fall in prices of agricultural commodities. This accounts for the persistent balance of payments' problem. Economists advocate industrialization as the solution to this problem so that production can be diversified. Due to this argument many of the country's industries are being renovated and new ones developed. However, old methods of waste disposal are still being used, among them the discharge of untreated effluents into water bodies.

The existing legislation is inadequate. The only Ugandan legislation on waste disposal requires that local authorities ensure cleanliness in their areas of jurisdiction. Therefore, these authorities see their task solely as removing wastes from the vicinity of the population of their area; what happens to the environment lies beyond their competence and even concern.

Consequently, Uganda is polluted by wastes because of inadequate or inappropriate methods of waste disposal. The government is therefore spending a lot of money to provide human and veterinary drugs, water treatment chemicals and other anti-pollution agents. Due to poor storage facilities, some of these chemicals, including medicines, become harmful to human beings and animals. Problems of disposal are not uncommon. For any type of waste, the main method of disposal is dumping on sites whose approval is not based on any environmental considerations.

Given the background above, it was concluded that waste management in Uganda is inadequate. A study has therefore been proposed to formulate a proper waste management strategy and the Department of Environment is moving fast to develop the necessary legislation for environmental protection generally and waste management in particular.

OBJECTIVES OF THE STUDY

- To identify and quantify the wastes produced;
- to document methods of safe disposal for the various wastes;
- to recommend an administrative structure and define responsibilities in the cycling of wastes;
- to formulate the criteria for approval of sites and facilities, within the framework of appropriate environmental legislation;
- to formulate standards to make wastes from generators acceptable for safe disposal;
- to make decision-makers appreciate the importance of a proper waste management strategy and to be able to invest in it by providing infrastructure and support services.

PRELIMINARY STUDY

A preliminary study has indicated that a variety of wastes are dumped at sites, designated by the local authority, sprayed with insecticides and covered with soil. Incineration and waste treatment are used only on a very small scale.

CONSTRAINTS

Little has been done to improve the situation described above. The government of Uganda established the Department of Environment only in 1986. Funding and finding skilled personnel are major problems. Much training is still required, as is relevant literature and other support measures, including IRPTC publications and links to its special files. In short, in the area of waste management, Uganda needs a great deal of UNEP help.

Chapter Twenty-Nine

The Global Environmental Monitoring System (GEMS)

MICHAEL GWYNNE
Former Director, GEMS Programme Activity Centre, UNEP

INTRODUCTION

Why is there need for a global environmental monitoring system?

Within the narrow band we call the biosphere exists all life as we know it. What other life there may be in the universe exists, if at all, light years away. This narrow band of life is all we have—and we, as the dominant species of its confines, are responsible for maintaining it. Consequently, we must know its workings and its composition—and the nature of changes in both—in short, the bases for environmental management of any kind.

The Action Plan for the Human Environment adopted by the Stockholm Conference of 1972 contained the seed of Earthwatch. It postulated three areas of work: environmental assessment, environmental management and supporting measures. Today, GEMS, the Global Environmental Monitoring System, lies at the heart of Earthwatch. Its tasks are the gathering and interpretation of information. It seeks to make comprehensive assessments of major environmental processes and thus provide the data necessary for rational management of natural resources and the environment. It also seeks to provide early warning of environmental changes from the data it monitors.

These enormous tasks are carried out in 142 countries in which UNEP works with other bodies of the United Nations' system, notably the World Health Organization (WHO), the United Nations Food and Agricultural Organization (FAO); the world's leading scientific NGOs; and governments themselves. This cooperation gives us a fair idea of the global picture. From that, national pictures, however blurred, may be drawn. Only when sufficient GEMS' activities can be carried out in every country—and when all the data gathered through vastly different techniques and types of instrumentation in different parts of the world using disparate data forms are harmonized—will many national pictures become as sharp as they should be. At present, the largest gaps lie in the developing countries.

Monitoring is assessment over time. Fundamentally, the process begins with identifying a problem and establishing its priority. Monitoring is constant. At some point, discernible trends begin emerging—the basis for prediction. At that point, one can begin to develop management plans. The effectiveness of these plans can be judged by further monitoring through a series of feedback loops. If this is not done—or if the initial base figures are inaccurate—these plans can go wildly wrong, creating not only environmental but economic disaster.

To monitor the natural world, whose major components and processes are constantly inextricably intertwined, GEMS falsifies it, dividing it into four major areas of interest, which, of course, overlap: first, *atmosphere and climate*; second, *environmental pollutants*; third, *Earth's renewable resources*; and, fourth *environmental data*:

ENVIRONMENTAL DATA

To start first with the last, GEMS takes raw data and processes it to manageable and meaningful proportions, then presents it so that it reveals major trends—against which, for instance, management strategies can be tested. Much of this is done by our Monitoring and Assessment Research Centre (MARC) in London which publishes UNEP's biennial *Environmental Data Report*. Its 10 major sections cover environmental pollution, climate, natural resources, population, settlements, human health, energy, transport, wastes, natural disasters and international cooperation. MARC also deals with the raw stuff of obtaining data, an activity which cannot be overemphasized. It publishes research reports on monitoring and assessment techniques. It also runs at least one major workshop each year, training personnel from developing countries in low-cost monitoring and assessment methodologies and in establishing national databases.

ATMOSPHERE AND CLIMATE

Climate is the driving force of the biosphere. It is now changing significantly because human activity has produced vast quantities of materials that are beginning to transform atmospheric composition dramatically. The "greenhouse effect" is basically an alteration of Earth's re-radiation of the radiation it has received from the Sun. This re-radiation, whose wave-lengths differ from that which we receive, has difficulty penetrating a layer of atmospheric gases—primarily carbon dioxide produced by industry and power generation—but also chlorofluorocarbons (CFCs) used in coolants, fire extinguishers and solvents; and methane and nitrous oxide arising from agricultural activity. Much of the re-radiation is bounced back in a constant series of attempts to pass through this "green" layer. Trapped, this re-radiation is heating the Earth's surface and changing that powerhouse that drives the planetary atmospheric circulation pattern. Different parts of the globe are growing warmer or cooler; different patterns of rainfall are developing; sea level is rising because increased heat expands all volumes, including that of the ocean.

Monitoring of the oceans as well as the atmosphere is critical to the study of climate change. UNEP's Ocean and Coastal Area Programme Activity Centre is working with the Intergovernmental Oceanographic Commission (IOC) of UNESCO in a programme of climate-related monitoring that will, we hope, eventually enable us to answer such vital questions as how much rain falls into the oceans and how much carbon dioxide is absorbed by the oceans, as well as why such absorption takes place. WMO is a close working partner in this endeavour, as well as in atmospheric studies.

There has been a general heating of the Earth over time. But it has been erratic. Many periods of warming have levelled off. We don't know why. We can, however, chart these changes back many years, through samples of gases trapped in glaciers that can be dated fairly accurately. This kind of monitoring of the past helps us improve our present management of the environment. Looking back over the past 40 years, we can plot a steady rise in carbon dioxide.

Looking back more than a century to the tall chimneys built during the late industrial revolution—first in England to discharge factory pollutants towards the sky so that industrial townspeople would stop complaining about the quality of the air they breathed—we can begin to chart the rise of long-range transport of airborne pollutants. These wastes, especially nitrogen and sulphur oxides, came down as acid rain in Scandinavia and Russia. From time to time, it has been palpably acid, killing crops and aquatic life. Its best-known damage has been inflicted on forests. GEMS monitoring of this scourge led to the 1979 Convention on Long-Range Transboundary Air Pollution, just as its data contributed to the elaboration of the 1992 Convention on Climate Change.

Work on the phenomenon of acid rain is far from finished. Though the problem was identified in Europe, we know that it affects North America, parts of Africa and South America and that it is widespread in Asia. We know, too, that in addition to its direct damage to plants, it will change the characteristics of soils and also the composition of the lower ozone level with a probable increase in ozone that will injure all kinds of tissue. (Restoring the stratospheric ozone lower down.) However, because far too few monitoring stations exist, notably in the developing countries, we have little idea of how extensive acid rain actually is.

ENVIRONMENTAL POLLUTANTS

This brings us to the second area of GEMS' concern: environmental pollutants and their effect on people. In close cooperation with FAO and WHO, GEMS monitors contaminants of air, especially in urban areas, freshwater and food. The urban focus stems from the fact that by the year 2000, most people will be living in some kind of urban environment. There are now 22 mega-cities—urban agglomerations with populations of 10 million or more; this figure, even now, is low. Not only do these cities have an environment of their own; they influence the environment of everything around them—which, in turn, impacts on them. This, as well as many other factors, calls for a constant revision of the list of pollutants being monitored.

One of the problems of looking at people is trying to monitor *total* exposure—the contaminants they absorb through breathing, drinking and eating, as well as through their skin. This concept was largely scoffed at in 1984, when WHO and UNEP set up the Human Exposure Assessment Locations (HEALS) to collect data on the exposure of different population groups from different sources, but its value in safeguarding human health has now been realized, particularly for certain age levels and job categories. Similarly, the UNEP/WHO programme for monitoring nuclear radiation, which began before the 1974 Test Ban Treaty, languished for years until the Chernobyl accident. However, with the establishment of the Global Environmental Radiation Monitoring Network (GERMON) in 1987, 40 countries, half of them in the southern world, began participating in the collection of data on radioactivity in air, precipitation, milk and other substances.

All this leads us to the concept of *integrated monitoring*. Repeated measurements of the same variables in the different environmental compartments—such as air, water, food and the oceans—allow us to trace the path of substances as they move from the air to the sea, the sea to the food chain and from the food chain to human beings. This is vital to environmental management and has taken a significant step forward towards an integrated methodology for recording and analysing the results of all kinds of environmental monitoring through GEMS' Global Resource Information Database (GRID), established in 1985. GRID has become a keystone in the bridge between monitoring and assessment, on the one hand, and, on the other, environmental management.

EARTH'S RENEWABLE RESOURCES

The third broad area of GEMS' concern is the renewable resources of the planet, primarily its diversity of life. Through the World Conservation Monitoring Centre, a joint effort of UNEP, the International Union for Conservation of Nature and Natural Resources (IUCN) and the Worldwide Fund for Nature (WWF), GEMS is concentrating on endangered species and their habitats in an effort to conserve the world-wide gene pool. We need to be able to determine species' content and their changes over time. However, below the vertebrate level, we don't yet have the methodology to do so.

The study of biological diversity brings us back to the necessity of improving monitoring and assessment at the national level because the study of species, notably small species, vertebrate or non-vertebrate, is best carried out on the ground, i.e. at the national and local level. As indicated earlier, global data are drawn largely from satellite imaging and aerial reconnaissance. What is missing in so many parts of the world today is the picture on the ground, *ground truth*. In short, throughout many parts of the developing world today, the essential component for ecological monitoring—and therefore environmental management—is simply lacking. We cannot hope to save what is probably our most precious renewable resource unless those ground gaps are filled.

DISCUSSION

Question: How can UNEP help developing countries to establish monitoring stations?

Answer: A number of centres are being set up through programmes conducted for GEMS by the European Office of the United Nations Institute for Training and Research (UNITAR). The first step is fielding a country mission to compile information on national needs for resource monitoring and assessment capabilities. After this, resource managers are trained in image processing and Geographical Information System (GIS) techniques, as well as related subjects. When these trainees return to their national institution, they are provided, on long-term loan, with the hardware and software needed to establish the nucleus of a GRID-compatible monitoring and assistance centre. Follow-up missions from GRID analysts provide further assistance and help design and carry out the first national pilot project. Most countries have far more environmental information than expected. The task is to concentrate this information so that it is readily accessible.

Unfortunately, a frequent problem in this process is brain drain from the developing countries, trained people who leave after having received UNEP and UNITAR assistance.

Question: What measures does UNEP take to treat disclosure of the security aspects of satellite images?

Answer: This is all public-domain material and can be made commercially available. We buy, when we can afford it, from commercial sources. We normally buy the tapes and have them processed. However, we do not hold vast resources of satellite information because its storage is too costly.

Question: Is there any legislation that can be established to prosecute the sellers and users of hazardous chemicals like DDT?

Answer: There are two sides to the use of toxic chemicals. Their negative environmental impact is well known, but they may be very effective in the treatment of a particular problem, such as using DDT against the tsetse fly. From that point of view, they may be economically worthwhile and may also save lives.

Question: Why is Antarctica important to human life?

Answer: Antarctica is the coldest area of the globe. Its land and inland water mass causes temperature differences that influence atmospheric circulation. We need to know more about these temperature effects, as well as why the largest ozone hole exists above Antarctica. The atmospheric fluctuation patterns caused by Antarctica affect the entire planet. No one country or area is isolated from any other. Antarctica is fundamental to Earth's climate equilibrium and therefore *must* be protected.

Chapter Thirty

The Global Resource Information Database (GRID)

NORBERTO FERNANDEZ
Manager, GRID-Nairobi

INTRODUCTION

GRID, the Global Resource Information Database, was established within the framework of GEMS in 1985 to form a bridge between environmental monitoring, on the one hand, and, on the other, environmental management, especially at the national level. It has become an integral element of UNEP's Earthwatch and adds a spatial dimension to ecological thinking because it is based on geographic information systems (GIS).

GIS

GIS integrates data from a variety of sources, such as satellite records, maps and surveys, and links to geographical locations a wide range of attributes, such as population density, land use, soil degradation severity and water resources to produce tables, graphs and a number of map-like materials to respond to the needs of scientists and planners. Information is stored in two databases: one devoted to spatial areas based on political or natural boundaries; the other to attributes such as those just indicated.

Thus, GIS can be used to superimpose different data sets on a particular geographical area so as to enable users to visualize, quantify and model the interaction of a number of different parameters such as topography, soil, vegetation and weather patterns. This also permits the creation of "What if?" scenarios for analysis, planning and management.

One such highly practical exercise was carried out in Uganda in 1987 to determine the best possible locations for growing a wide spectrum of crops. Together, GRID specialists and Ugandan experts built up an environmental database for that country containing datasets on land, climate and infrastructure. That system was then used to generate a soil productivity map that included such elements as soil texture, depth, acidity, fertility and workability. A comparable soil erosion map was evolved from the erosiveness of the country's soils in relation to rainfall, slope, pressure from land use and human population density. Then the team added a "What if?" factor—in this case a temperature increase of two degrees Celsius, which, in itself, expressed concern with the effects of the anticipated phenomenon of global warming. The results, for coffee cultivation alone, one of Uganda's chief cash crops, were startling.

This is but one example of the use of GIS technology in helping planners faced with development problems. Other applications have been used in a wide spectrum of environmental studies, including the following:

- an estimate of the number of elephants left in the whole of sub-Saharan Africa, based on surveys covering smaller areas;
- an investigation of a tick-borne disease that identified areas in Africa potentially vulnerable to East Coast Fever;
- an evaluation of the suitability of coastal sites in Costa Rica for the development of aquaculture;
- an evaluation of new methods of mapping and monitoring the world's tropical forests, using satellite data.

THE GRID NETWORK

GRID currently operates through seven centres, each of which serves as an archive of regional and sectoral datasets in GRID formats as well as a potential repository of new datasets for the system and a laboratory for GIS and other information-processing technology. Each centre also distributes data on standard media and provides support to new national GRID-compatible centres such as that of Uganda. Nairobi and Geneva GRID are funded by UNEP. The Bangkok Regional GRID Centre is supported by UNEP, the United States' National Aeronautics and Space Administration (NASA) and the Asian Institute of Technology. The three other GRID Centres are supported primarily by national funding and are located in Sioux Falls, USA; Arundel, Norway; and Warsaw, Poland. Others envisioned in the near future could be in Germany, Russia, India, Japan, Jamaica, Mexico, Brazil and the south Pacific.

Chapter Thirty-One

INFOTERRA, the international environmental information system

WO YEN LEE
Former Director, INFOTERRA Programme Activity Centre, UNEP

INTRODUCTION

INFOTERRA is an international system that facilitates the flow of environmental information within and between countries. Like GEMS, it lies at the heart of Earthwatch, growing out of a recommendation of the Stockholm Conference. The INFOTERRA concept was developed from 1973 to 1974 and preparatory work was carried out during 1975 and 1976. The system became fully operational in January 1977 with the participation of a dozen countries. At present, 138 countries have designated national focal points that cover over 99 per cent of the world's population. Over 6500 sources are listed in the *INFOTERRA Directory*, which links a quarter of a million experts to the network. A cumulative total of over 210 000 queries have been processed for users in 114 countries during the last 14 years.

INFOTERRA was designed as a decentralized system, operating through government-designated national focal points. This decentralized structure has proved to be the least costly means of facilitating information exchange, as well as the most effective in fostering the improvement of national information systems by governments—or their creation in the many cases in which none existed. The fundamental aspects of INFOTERRA have been defined as decentralization; provision of information to decision-makers through the sources registered in the *INFOTERRA Directory* (national or international); facilitation of the exchange of environmental information through access to environmental databases; promotion of awareness of the role and importance of information in environmental decision-making; and stimulation of the development of national environmental information systems.

STRUCTURE OF THE SYSTEM

The INFOTERRA information system consists of five components: national focal points (NFPs), sources of information, Special Sectoral Sources, Regional Service Centres and the INFOTERRA Programme Activity Centre.

National focal points

The INFOTERRA national focal points are the key elements in the network, as they provide a sort of national *Who's Who* of the country's environmental expertise and select the best sources for inclusion in the *International Directory*, which is one of the system's main tools. They are also the first points of contact with the users. Many of them, especially in developed countries, consider the informal and relatively easy access to national environmental information systems and their dialogue with them one of the greatest benefits they derive from their participation in INFOTERRA. Others, especially those in developing countries, consider their participation in the INFOTERRA system a cheap and effective way of gaining access to modern science and technology to solve their environmental problems. After receiving INFOTERRA training, the NFPs are expected to perform a number of functions, including the registration of sources and the processing of queries, and to serve as the INFOTERRA linkage in the country. Most of the NFPs are located in the information branch of the central environmental department, to which people naturally turn when they have an environmental query.

Sources of information

The community of sources listed in the International Directory forms the backbone of the INFOTERRA database. Judging from the feedback obtained from a routine users' satisfaction survey, the sources do in most cases provide useful information. The 6500 institutions registered from 106 countries represent some 250 000 experts who share their expertise in over 1300 priority subject areas with anyone who needs to draw upon environmental experience in these fields. Through this network, 2.5 million publications on various environmental subjects have been made available to users, in most cases free of charge.

Special Sectoral Sources

Special Sectoral Sources are centres of excellence in selected environmental priority sectors. They are world-renowned organizations that can provide comprehensive, authoritative information in their own areas. These organizations are contracted by UNEP to provide substantive information, at a nominal charge, to users anywhere in the world. The selection of the queries forwarded to receive this additional help is based on the nature and the origin of the query, priority being given to users from the governments of developing countries—largely policymakers, scientists and engineers. For example, in the United Kingdom the Environmental Legislation Information System in the field of environmental legislation, and the Harwell Laboratory in the field of waste treatment, serve as Special Sectoral Sources.

Regional Service Centres

Given the similarity of many environmental problems within a given region or sub-region, regional centres for environmental information have been appointed as Regional Service Centres. This permits services such as computer search facilities, training, promotion, the provision of substantive information to be provided more economically or in a more professional manner. So far nine centres have been established to serve nine developing regions, including Southeastern Asia, Southern Asia, Northern Africa, Western Asia, Eastern Africa, Western Africa, Latin America, the Caribbean and East European countries.

PROGRAMME ACTIVITY CENTRE

The Programme Activity Centre was set up as an internal project of UNEP to coordinate the network. Its major tasks, in line with the catalytic and coordinating role of UNEP, have been assisting governments in establishing and developing INFOTERRA national focal points; providing the necessary training to national focal point staff, especially in developing countries; providing system tools and improving system operations; assisting with directory searches and developing model publicity materials. Coordination was achieved through designing standardized operational procedures and terminology as well as through publishing the *INFOTERRA International Directory*. Given the decentralized structure of the INFOTERRA network, however, INFOTERRA operations in individual participating countries depend almost entirely on the efforts of their national focal points and the government support they receive.

THE INFOTERRA SYSTEM TOOLS

INFOTERRA was conceived as a system of the utmost simplicity. It was designed to yield valid results with minimum of professional information system expertise. Its principal tool, the *International Directory of Sources* has undergone a number of format changes, while many attempts have been made to reduce the volume of this very substantial publication. INFOTERRA developed its own software and procedures for the compilation of its database, and adopted the CDS/ISIS system for the micro-computer version of its database.

Another significant tool of the System is its terminology. The entire vocabulary contains some 1300 priority subject areas. INFOTERRA operational procedures are described in detail in an *Operational Manual*, which is published in four languages and is made available to all national focal points.

An electronic mail system linking many of the INFOTERRA network partners was initiated in November 1988. Together with other means of modern communications, such as fax, the typical turnover time for a query–response has been reduced from weeks to days.

THE IMPACT OF INFOTERRA

Query–response services

As indicated above, through the query–response services of INFOTERRA, solutions to environmental problems and other substantive information have been provided information seekers in 114 countries for over 210 000 queries during the last 14 years and have led in many cases to more effective management decisions and improvements in environmental quality. The main subject areas of enquiry have been pollution control, waste treatment, chemical and biological agents, technology and industry, and management and planning.

During 1990, through these information services, several countries managed to resolve problems associated with trans-boundary movements of hazardous wastes; established national legislation on various aspects of the environment, including the marine environment; cleared accidental spillage of potentially toxic chemicals; managed to contain the invasion of swarms of locusts; improved the efficiency of energy use and production; and advanced their skills for the treatment of industrial effluents and discharges.

INFOTERRA'S CATALYTIC ROLE

In accordance with UNEP's mandate for catalysis, INFOTERRA has promoted the establishment of national environmental information systems; assisted in setting up international information systems of relevance to the environment; heightened environmental awareness whenever appropriate; and advanced the participation of developing countries in the international exchange of environmental experience. Notable examples of national environmental information systems are those in Brazil, China, Colombia and India. These and other countries are establishing comprehensive national environmental information systems as part of a broader national information effort, with the encouragement and support of INFOTERRA. For the majority of INFOTERRA's partner countries, their participation in INFOTERRA activities has resulted in strengthening their national infrastructure for handling environmental information.

INFOTERRA assisted several international environmental information systems during their formative stages, notably the CLENR and ASFIS systems of the Food and Agriculture Organization of the United Nations (FAO), MEDI of the Intergovernmental Oceanographic Commission of UNESCO, ELIS of IUCN and ICSTI for the East European countries. INFOTERRA has maintained very close links with all information systems within the United Nations, organizing its services to adapt to the evolving needs of developing as well as developed countries.

DISCUSSION

Question: Can INFOTERRA provide practical help when an emergency erupts in a developing country that does not have the infrastructure to cope with it?

Table 1 Impact of INFOTERRA: some success stories

Country	Success Story
Belize	Rejected the offer of a used-oil treatment factory from a multinational company, based on environmental impact information received from INFOTERRA.
China	Developed national environmental monitoring network, reduced urban noise level, developed national safety colouring code, regained contact with world-wide learned societies, received a number of technological details in various fields.
Gambia	Improved large-scale rice plantation projects and prevented hippos from damaging the paddies.
Kuwait	Developed coastal areas based on environmental guidelines.
Malaysia	Rejected the offer of a TiO_2 manufacturing factory by a multinational company, made better use of rice husks, improved the efficiency of spraying of 20 named pesticides.
Oman	Developed oil-spill contingency plan, appropriate methods of disposal of solid municipal wastes and rectified the dangers of hydrogen sulfide-contaminated wells.
Samoa	Rejected the offer of a used-oil treatment factory from a multinational company.

Answer: At present INFOTERRA is essentially an information system. However, the United Nations Centre for Urgent Environmental Assistance is soon to be established in Geneva. It will have experts on call, as well as the ability to provide financial assistance through the mobilization of donor aid.

Question: How long does it normally take to receive a response from INFOTERRA?
Answer: There is no standard time. It may range from one day to two weeks, depending on the question's complexity and the availability of experts to deal with it.

Question: Are National Focal Points individuals or institutions?
Answer: Offices, such as a national environmental protection bureau, serve as NFPs, so that when personnel are transferred, the continuity of the function can be ensured. If a government so requests, INFOTERRA can help train new staff.

Chapter Thirty-Two

Technology transfer, environmental auditing and environmental impact assessment

YUSUF J. AHMAD
Formerly, Special Adviser to the Executive Director, UNEP

INTRODUCTION

Divergent as these three subjects may seem at first glance, they are in fact very closely related. Moreover, all three are "over-arching" concerns. Like population, they transcend and unite conventional disciplines and sectors. All are essential to the proper management of our increasingly stressed planet.

TRANSFER OF TECHNOLOGY

The transfer of technology that is benign in its effects, efficient in its use of energy and amenable to the recycling of wastes is the key to integrating environment with development. It is also a critical point of entry to resolving global problems such as the depletion of the ozone layer, climate change and today's unprecedented destruction of biological diversity—dilemmas that simply cannot be addressed by any one country or group of countries. Quite rightly, the less-developed countries claim that if they enter into legal commitments to mitigate these problems, they must have legal guarantees of receiving the best possible technologies to fulfil these commitments, as well as the transfer of the financial resources to meet the incremental costs involved in reducing major types of pollution and other environmental damage. Many in the industrialized countries agree. However, these technologies, far from belonging to governments, are owned by industrial conglomerates, many of which are transnational, and all of which have spent enormous sums on research and development to produce alternatives to environmentally destructive processes. They, too, have a right to a fair return on their investment. Consequently, in one way or another, the less-developed countries need considerable concessional aid if they are to play their proper role in Earth's environmental drama—in short, financial transfers to support technology transfers.

UNEP significantly facilitated one such transaction during the elaboration of the

Montreal Protocol on Substances that Deplete the Ozone Layer. A wide spectrum of estimates was put forth for chemicals and equipment to reduce the problem, so that the shape of the cost curve became a contentious issue. UNEP maintained that this curve would be *concave*—that costs would rise at the beginning of the conversion process, then flatten into a plateau, and at last fall by 1997 because of the development of new technologies and substitutes for CFCs and halons, as well as the competition among the companies marketing these products. Even so, no agreement emerged on the figures. Therefore UNEP drew the attention of the contending experts away from figures and towards the establishment of a three-year roll-over fund—a proposal that led to consensus.

We must also bear in mind the changing picture of technology transfer during the last two decades. The developing countries no longer have to use the technologies given to them—often dumped—by the industrially advanced. The United States Office of Technology Assessment now estimates that no fewer than 140 new technologies emerge each year in biotechnology alone. Consequently, the less-developed countries can now choose from this immense proliferation in terms of their own needs. Moreover, they can now judge these new developments not only in the light of capital and labour requirements but according to their particular economic, social, cultural and environmental conditions. Also gone is the time when the less-developed countries viewed environmental protection as a luxury. In the course of their own development, they witnessed the burgeoning of cities like Mexico City, with its immense pollution problems, as well as tragedies like the Bhopal explosion. We now know that development and environmental protection are inseparable. What is needed is far greater knowledge of the available technologies to serve these ends. In this connection, UNEP has established an international Environmental Technology Centre in Japan to deal with, among other things, technology transfer. We also need to assess the true environmental costs of development and the environmental impacts of development projects.

ENVIRONMENTAL AUDITING AND RESOURCE ACCOUNTING

Many countries, especially those that are highly industrialized, are introducing environmental auditing and natural resource accounting to determine the state of a country's patrimony—its air, water, forests, soils and other elements of its legacy from nature. When one compares detailed projections of gross domestic product with those computed for natural resources, one usually finds that the country has been living off its capital.

The current system of national accounts was developed by the United Nations. They built on the work of the 1940s carried out by economists primarily concerned with countries emerging from the Great Depression and the subsequent havoc of World War II—usually by increasing government investment in various sectors of the economy. But these accounts did not include the value of natural resources, the country's basic capital. We now know that gross domestic product will not be augmented by the felling of a forest.

Work has only recently begun on the valuation of natural resources. UNEP, the United Nations Statistical Office and the World Bank are cooperating to develop this type of auditing. UNEP has undertaken three major case studies—one in Latin America, one in Asia and another in Africa to explore the feasibility of collecting the required data and to determine what organizational arrangements are necessary to this task. The United Nations Statistical Office is preparing models for "satellite accounts" that will reveal such factors as energy consumption and agricultural production. The World Bank is clarifying the conceptual issue of how to place a value on environmental *functions*. In the near future, all three bodies will pool their findings to revise the present system of national accounts. When this methodology is developed, gross domestic product figures will probably decrease. Some may even be negative.

This sort of reckoning must be carried out because it performs two essential tasks. First, it supplies a data-base in physical and monetary terms that enables planners to assess the interaction between environment and development—in short, to evaluate the environmental management of their country. Second, it determines the physical wealth of a country and thereby provides a key indicator of sustainable development. Every country, especially the less developed, needs this kind of comprehensive inventory. It must then deduct from its projected gross domestic product what economists call "defensive expenditures", the costs incurred by the wrongs done to the natural resource base and the environment. On the basis of this balance, true planning for sustainable development can begin.

ENVIRONMENTAL IMPACT ASSESSMENT

This type of planning calls for environmental impact assessment (EIA). Unlike conventional economics, in which the scarcity of one good or another coupled with its demand, determines its value, enviromental evaluation inheres in the interconnection among all the elements of a given ecosystem. It calls for a holistic approach where sectoral planning would be replaced by integrated planning, even if only to stem the environmental damage that has occurred since at least 1945. Otherwise, the principle of sustainable development is doomed—and, with it, our planet.

But we cannot ignore the difficulties and cost of environmental impact assessment. An EIA should be undertaken after the economic engineering feasibility study has been carried out. The EIA team should first draw up a comprehensive list of all the project's possible impacts, including social elements. From this comprehensive list, the team should cull a *manageable* list comprising some seven or eight impacts out of one hundred. Though these will differ from place to place and project to project, they should be determined by four criteria:

- the magnitude of the possible impacts;
- their ecological significance;
- their extent in space and time;

- the sensitivity or vulnerability of the elements in the country or physical area.

In developing countries, these sensitive elements usually include land and water. The manageable list should also reveal a threshold of possible impacts. If this threshold is correctly gauged, the projected costs of the project may well increase. If it is set too low, the EIA will probably be worse than useless.

The EIA should also be accompanied by a risk and uncertainty analysis. All development involves intervening in nature and we don't yet know the consequences of our interventions. We do know, however, that all have systemic interconnection. Some may be synergistic, the first and second being benign and the third harmful. Impacts can be irreversible—and usually are. For this and many other reasons, the risk analysis should be made and kept close to the decision-making centre.

To increase its accuracy, as well as to reduce its costs, the EIA should be carried out by an interdisciplinary national team, *not* a foreign consultant. Foreigners usually cannot tap the local knowledge necessary to the validity of the assessment. Moreover, consulting local people creates the popular participation without which the project, however viable in physical or economic terms, may fail. Local newspapers and community groups that favour a particular environmental action or effect will tend to create political pressure for it and thus enhance the project's implementation.

In many developing countries, environmental impact assessments still tend to be perceived as delaying tactics for development. But in view of all the concerns discussed above, the EIA is also a powerful tool for improving project design and therefore its success in all dimensions.

DISCUSSION

Transfer of technology

Question: Could developed countries foster transfers of technology by granting private entrepreneurs such incentives as tax exemptions and lower fees for patent licensing?

Answer: In view of the magnitude of the problems we now face, such approaches may be insufficient. The difficult now lies in obtaining the best possible technology, which is heavily guarded. A more systematic approach is necessary.

Question: How would a clearing-house work?

Answer: UNEP envisages cooperating with partners to establish clearing-houses geared to specific topics. For example, a clearing-house for energy technology has been established in Denmark. Clearing-houses play the role of honest brokers. When they receive requests from developing countries on specific plans to build industrial plants, they circulate this information to developed countries, asking them to identify the most appropriate technology. These responses are then passed on to developing countries.

Question: Given the difficulties of technology transfer, shouldn't developing countries try to develop their own technology, which may also be more appropriate?
Answer: Science and technology are now progressing so rapidly that there is no alternative but to search among modern technologies to achieve effective and efficient results. There are great difficulties in transferring indigenous technologies from one developing country to another because specific conditions may be very different.

Question: Many technological breakthroughs have been made by people from developing countries in the laboratories and research institutions of developed countries. How can this brain drain be reversed?
Answer: Scientists from developing countries are often unable to find in their native lands the climate, opportunities and infrastructure to pursue their research. Perhaps international organizations should foster the creation of better research conditions to persuade their scientists to stay. Then there is the question of research facilities.
Comment: The United States Agency for International Development has proposed technical cooperation with developing countries by establishing training centres in developing countries in which local people could be trained in environment-related technology.

Question: What is UNEP doing to reduce pollution problems that are worsening in developing countries because of their inability to import appropriate technology?
Answer: One way is inducing polluters to pay for their pollution through a market-based incentive system.

Question: Although a Ghanaian herbal clinic has claimed to have identified a cure for AIDS, there have been attempts to take the technology out of the country and improve it elsewhere rather than support its development in Ghana. How can such risks be averted?
Answer: This is an issue of protection of technology and proprietary rights that stems from appropriate legislation and adherence to international intellectual property rights agreements.

Environmental auditing and resource accounting

Question: What is the difference between environmental economics and environmental accounting?
Answer: Environmental economics is a new science that has grown out of the recognition that since natural resources are becoming increasingly scarce, this scarcity should be reflected in market values. Another function of environmental economics is to quantify the market failure that results from nature's decreasing capacity to absorb wastes. Environmental accounting is an instrument of environmental economics designed to avert market failures by calculating appropriate market values for environmental goods and services.

Question: Which organizations are responsible for environmental accounting and auditing?

Answer: The World Bank, the United Nations Statistical Office and UNEP began work on this subject in 1984. One result is a recent report by the World Bank entitled *Environmental Accounting for Sustainable Development*.

Question: How can natural resource accounting be integrated into the system of national accounts when the latter reflect market prices, whereas the former do not?

Answer: Given the great difficulties of accountants in integrating non-homogeneous concepts, at this stage they should complement traditional national accounts with satellite environmental accounts.

Environmental impact assessment

Question: What kinds of projects require EIA studies?

Answer: This should be decided by the government in terms of the sensitivity of specific issues.

Question: In the absence of legislation concerning EIAs, should the project developer carry out the study or should the government institute a body for such assessments?

Answer: An appropriate government body must establish the terms of reference for the project and decide what is needed. The government must not depend on the developer in this respect.

Question: As EIA studies may take a year or more, how can the problem of timing *vis-à-vis* the budgeting process be resolved?

Answer: Procedures should be developed to make the scope and timing of the EIA proportional to the relevance of environmental issues. Time and money should not be wasted on EIAs concerned with marginal environmental issues.

Question: In carrying out EIAs, local experts can be pressured officially or politically. How can one be sure of their objectivity?

Answer: Foreign experts may not be altogether independent in their views either. The best approach is to form a team of experts and to give them the highest autonomy.

Chapter Thirty-Three

Pollution: its causes and control in industrialized countries: the case of Japan

TAKAHIKO HIRAISHI
Coordinator, Support Measures, UNEP

INTRODUCTION

An essential prior condition of the transfer of technology to developing nations is the transfer of the history of the frequently tragic mistakes of the industrialized countries. That of Japan is probably best represented by the Minimata tragedy, through the specific case of a mother and her child, victims of the debilitating, incurable and now infamous "Minimata disease", poisoning by a by-product of methyl mercury used by a factory in the small coastal fishing town of Minimata— an industrial plant that had been the glory of the little port, whose chief source of protein was fish. The contaminant began being discharged into Minimata Bay during the 1950s and entered the marine food chain. Only in 1968 was it detected and declared the cause of the disease, which can be transmitted through the placenta of a pregnant mother. By that time, some 2000 people had been stricken. Of these, 1000 have now died, but, despite the closure of the factory, the driving development engine of the area, the Japanese health authorities estimate that some 4000 potential victims remain.

The company still pays approximately US $20 million per year to those debilitated by Minimata disease. Mainly because of this, the factory is closed, and heavily supported by the Japanese authorities. Moreover, as Minimata Bay has been closed to fisheries, the company compensates the fishermen of the area for their lethal catch. When one notes that if only 1% of the company's profits during the 1950s had been devoted to the prevention of pollution and technology for its control, one realizes how staggering the losses have been in economic as well as human terms. Further, this particular case of environment-related disease is only one of so many in Japan alone, a country that pays over US $700 million annually (1990 figures) solely in pollution compensation.

STANDARD SETTING AND IMPLEMENTATION

The setting of pollution standards, as well as the legal penalties for violating them, is a thorny area. Japan uses some 100 000 industrial chemicals, most of which penetrate the environment at some point or another. No one can really foresee what health and economic effects any of them may have in the long term. Historically, local governments have always been the first to identify a pollution problem and often to initiate pollution control or prevention measures. National action is always belated.

Japan's Environment Agency was established in 1971 after a virtual explosion of local environmental pollution problems in 1970. It immediately adopted provisions similar to the United States' Clean Air Act, for car exhaust regulations, and implemented them even before the USA did. It quickly began monitoring the path, magnitude and possible impacts of the numerous industrial chemicals referred to above, certainly those that had been detected by local authorities, often alerted by the public. The Agency is a regulatory and coordinating body whose major operational thrust is the conservation of nature. Sectoral ministries had been implementing the pollution regulations that apply to their various activities. Standards must, of course, be developed scientifically, but they must be tempered, like the penalties for their violation, by what is deemed politically reasonable. Their application requires time. It also calls for public support and participation, which, in turn, call for public information and awareness programmes.

Fortunately, Japan is now an environmentally conscious country, at least in so far as industrial pollution is concerned. Although standards are set by the central government, the brunt of the responsibility for their implementation is actually and eagerly borne by the local authorities, which may impose far more stringent norms than those issued nationally. Moreover, the local authorities are twofold in nature: the prosecutor's office and the police, between which an interactive competition often arises. Consequently, to satisfy both for public relations' purposes, an enterprise that wishes to establish itself in the locality or to expand its activities there may well go beyond the stringent local standards and participate in the long-term developmental planning of the entire area.

How far has Japan come in its clean-up efforts? Certainly, its air quality has vastly improved with dramatic reductions of sulphur dioxide, nitrous oxide and carbon monoxide in particular. The country operates almost 1300 fluegas desulphurization units, as compared with some 300 in other industrialized countries. This figure should not be taken as an indication of pride, but rather an index of the amount of pollution generated by the country's rapid economic expansion after World War II. Similarly, Japan, unlike many other industrialized countries, has constructed denitrification facilities. Let it not be assumed, however, that Japan is even relatively free of pollution. Apart from industrial problems, to which we will return in a moment, much water pollution remains, particularly from municipal sources, in part because the disposal of household waste water cannot yet be regulated and in part because only 40 per cent of the country is served by sewage treatment plants, despite heavy investment in such construction. Most of Japan's lake pollution also stems from problems in the disposal and treatment of waste water.

INDUSTRY'S PARTICIPATION IN POLLUTION CONTROL

After strong initial protests in the early 1970s, Japanese industry began to launch significant research and development efforts in pollution control—not only because of its fears of Minimata-like catastrophes, but because the government provided financial incentives for such activities, notably in the form of tax reductions and low-interest loans for the installations of pollution control devices. By 1974, Japanese industry was investing 16 per cent of its profits in pollution control. This has now diminished to approximately 4 per cent, partly because a significant amount of Japanese pollution control investment had been completed by the early 1980s.

The types of subsidies offered by the government may well be viewed as a violation of the "Polluter Pays Principle", which stipulates that industry should pay all pollution costs, including those of prevention. However, for short-term development, the Japanese government considers such subsidies justifiable. Government incentives resulted, for instance, in energy and resource economies—crucial in a country dependent on imported oil—as well as car engine improvements and other inventions.

THE TRANSFER OF TECHNOLOGY

The year 1992 witnessed the establishment of the UNEP International Environmental Technology Centre in Japan. Its major functions are training, the dissemination of information, research into the modalities of technology transfer and consultancy services.

It must be remembered, however, that private companies rather than governments have developed the technologies that industrializing countries seek. Consequently, there cannot be free access to technologies. Even in the case of the Montreal Protocol on Substances that Deplete the Ozone Layer, the provisions concerning technology transfer are very vague, precisely because international financing is required to make the pricing of technology transfer reasonable. This must be done in other cases as well.

Even if the patent system could be changed, the transfer of technology would not be significantly expedited. Without know-how and industrial investment, a patent is useless. Today, much Japanese production is carried out abroad. This development arose from the series of oil crises that began in the 1970s, during which Japanese manufacturing enterprises relocated to other parts of Asia in a process of redistributing heavy industry. This may well be considered the export of pollution. None the less, international investment is the best conveyer of technology transfer, in no small measure because that transfer includes pollution control technology.

DISCUSSION

Question: Who pays the compensation costs to pollution victims, the government or the private companies?

Answer: The basic principle is that polluters pay compensation. Air polluting industry pays 80 per cent, while 20 per cent of air pollution compensation costs are paid out of automobile taxes, which come directly from users to the environmental agency.

Question: In reference to the Minimata case, if 4000 potential victims are still anticipated, does this mean that mercury effects continue even after clean-up activities have been undertaken?

Answer: These pollution activities stopped in 1968. Pollution removal activities continued at the port where contaminated bottom sludges had to be dredged and disposed of in sealed containers. Although no known patients were detected, some symptoms were noticed in ageing people. However, these symptoms could not be completely differentiated from those related to other diseases and ageing. Still, we do not know enough about the transmission of the disease or when it may become manifest to dismiss a projection of many, many potential cases.

Chapter Thirty-Four

Environmental management and economic development in the Seychelles

JOSEPH E. L. FAURE

Assistant Director, Parks and Gardens, Department of the Environment, Seychelles

The Seychelles is basically a series of small and fragile island ecosystems. Its island environment is not only a unique asset of truly rare beauty; it is also highly vulnerable to disruption and degradation. Ill-conceived development initiatives that may have little noticeable effect on large continental land areas can cause permanent and irreparable damage to island ecosystems. To pollute or mismanage them would seriously undermine Seychelles' economic development prospects.

As inhabitable and arable land is so limited, the nation must manage on and with the little space it has. The country simply cannot afford the consequences of environmental degradation and unsustainable development anywhere in and around our small island homes. The government has therefore committed itself to an explicit policy of sustainable economic development and made it the principal theme and thrust for its 1990–1994 National Development Plan (SNDP).

This commitment is represented by the following major policy goals:

- *to protect the health and quality of life of all Seychellois* by setting standards and enforcing health and environmental quality standards for air, water, noise and marine pollution, the disposal of solid wastes and sewerage, the use of chemicals and worker health and safety;
- *to ensure that future economic development proceeds on an equitable and sustainable basis* in order to meet the needs and aspirations of the present generation without undermining the possibilities of future generations by making sustainable use of renewable resources and efficient use of non-renewable energy and mineral resources;
- *to preserve natural heritage and biological diversity* by protecting the unique or endangered species, ecosystems and ecological processes and observing the principle of optimum sustainable yield for living natural resources and ecosystems;

- *to improve decision-making, laws and the institutional framework for sustainable development* by undertaking prior environmental assessments of major new activities, by making the sectoral and economic agencies directly responsible and accountable for ensuring that their policies, programmes and budgets support development that is sustainable and by strengthening the legal and institutional framework for environmental protection and natural resources management;
- *to increase public information and understanding of the essential linkages between environment and development* by regularly publishing relevant data and information on changes in the state of the environment and natural resource base and by expanding environmental education and training programmes;
- *to strengthen international law and reinforce international cooperation* on economic and environmental matters by ratifying and implementing relevant conventions and expanding our participation and contributions for relevant sub-regional, regional and global organizations' programmes and conventions.

Meeting these policy goals requires some reorientation of present key policies, strategies and programmes of many key government agencies. Realizing these goals also requires the active participation of the population, as well as government consultation with and support of the private sector. The transition will take time. The 1990–1994 National Development Plan provides the starting point for this necessary period.

An attempt has been made to integrate environmental and natural resource management issues into all the relevant economic strategies and policies, the intention being to facilitate the incorporation of environmental management into economic decision-making, which is a prerequisite for sustainable development.

The environmental and natural resource management projects' and sub-projects' components in the NDP have been consolidated and expanded to form the Seychelles' Environmental Management Plan (EMPS-90) that is a more detailed and operational extension of the central core of environment-related projects in the 1990–1994 NDP. Further, the EMPS-90 includes the crucial environmental protection measures that are necessary for ensuring that all the other projects in the 1990–1994 NDP contribute to development that is economically and ecologically sustainable.

In an important way, the Seychelles is marketing the environment—and is thereby effectively earning foreign exchange to finance other social and economic projects. It is in this sense that the two plans form an integrated and complementary package and that the implementation of the EMPS is intrinsic to the success of the overall strategy set out in the NDP.

In summary, the EMPS is not expected to impose an unreasonable burden on institutional manpower or financial capacity in Seychelles. Because the EMPS is not a separate exercise, but an integrated part of the NDP, it will not be necessary to create additional systems to ensure proper implementation and monitoring. Only modest adaptations will be required to existing institutional mechanisms and planning procedures to ensure effective consultation, coordination and cooperation in implementing both Plans.

Chapter Thirty-Five

Thailand launches villager environment conservation project

YENRUDEE SUPUNWONG
Office of National Environment Board, Ministry of Science, Technology and Energy, Thailand

Designed to assess the potential for nation-wide application of villager-based environmental conservation and management activities, the initial pilot/feasibility phase of Thailand's Villager Environment Conservation Project ran from March 1991 until November 1992. The project is headquartered in Bangkok, with a sub-office in Kohn Kaen, a northeastern province, and operates in four zones spread over 10 provinces in the north, northeast, west and south of Thailand. Project assistance is provided to 58 key villages and reaches approximately 2500 households (over 12 500 people).

The project is funded by the Asian Development Bank and the Royal Thai government through the Office of the National Environment Board. Technical assistance, in the form of international advisers, is provided through the International Development Support Services of Australia.

The main thrust of the project is to raise the awareness of local communities about environmental issues and to encourage these communities to take actions to protect their environment. The programme aims to do this in a manner that does not take the initiative away from the local communities but strengthens their capacity to undertake self-reliant and sustainable activities consistent with regional and national environment protection objectives. The Office of the National Environment Board (ONEB), located within the Ministry of Science and Technology, was identified as the primary agency for project implementation and coordination.

The project is collaborative. It is an integration of technical responses, training and education, addressing environmental problems through training and non-formal education processes. In addition, it is a community-based project, since its success is based on the active involvement and commitment of the concerned rural communities.

Underlying the selection of project activities is the belief that focusing on environmental issues alone is not enough. A single theoretical approach is inappropriate for target rural populations. This concept supports the government's decision to link environmental issues directly and pragmatically to the future survival of individual village communities. Coordinated by the ONEB, the project is implemented through selected NGOs.

The choice of project activities is based on the understanding that environmental problems must be viewed in the context of the socio-economic problems of a significant proportion of the Thai population. Increasingly, problems cannot be addressed in isolation; conservation measures must be married to development objectives if the rural poor are to be converted to being effective natural resource managers.

For this reason, initial project activities focus on carrying out socio-economic and natural resource surveys and profiles of target communities, identifying and analysing key environmental problems in order to identify and evaluate appropriate and viable natural resource conservation and management techniques. Non-formal education, and community-based training, will use a number of sub-project activities as a means of raising environmental awareness and concern realistically and pragmatically. The project activities include the following project areas:

FARMING SYSTEMS

Existing farming systems centre on wet-season crop production. Unreliable rainfall, lack of capital and high debt levels have constrained intensification of existing systems. The objectives of this sub-project are to identify major constraints on increased crop production; to identify and evaluate additional cropping options to diversity and stabilize the existing farming system in an environmentally sustainable manner. Activities include cultivation of wet season legumes, alternate upland cropping options and dry season soy bean/mung bean production after rice cultivation in the lowland areas.

Stable farming systems will allow farmers to crop the same piece of land continuously by preventing soil erosion and by retaining soil fertility. They will also assist in stabilizing the village and returning certain lands to forests.

AGRICULTURAL DEVELOPMENT AND EXTENSION

The project's major activities are agricultural information, including recommendations on use of organic fertilizers; horticultural programmes to encourage the introduction of tree crops; seed exchange programmes; the introduction of the mung bean before rice planting in lowland areas; domestic livestock programmes, including buffalo banks; aquaculture programmes; and rice banks. These programmes are aimed at improving and stabilizing farm incomes. Agricultural extension is linked closely to the farming systems programme, making maximum usage of experienced NGO staff and dealing with experienced farmers.

COMMUNITY SELF-RELIANCE

It is project policy that all components emphasize community participation as a means of achieving an increasing level of community self-reliance. In order to ensure wide community participation, NGOs with deep grassroots links with local communities have been selected, and accurate data on the socio-economic and power structures is being collected for use in the formulation of approaches to each community. It is recognized that environmentally degrading behaviour is often a developmental problem and that strategies to reverse this behaviour must incorporate an understanding of the need to promote the overall social and economic development of these communities.

TRAINING

NGO staff and villagers are being taken through the process of understanding the need for gradual and phased community education on the problems and solutions to be adopted. They are being familiarized with the use of non-formal education methods, and materials are being developed progressively to support such training. The development of individualized plans of action incorporates the different and specifically relevant circumstances of each project location.

Project staff, together with NGO staff, are offering training in community, organization techniques; it is increasingly recognized that protection and conservation measures may not be feasible unless the local populations that are dependent upon the natural resources are adequately informed about the ecological functions of these resources, as well as the negative results of mismanagement.

Village leaders and resource persons are being trained and supported to carry out village-level community education and training in environmental awareness and various aspects of technical assistance designed to improve farming systems, so as to move towards an environmentally sustainable living pattern.

Many NGOs have been unsuccessful in their development efforts, not because their activities were inappropriate, but because the villagers were not sufficiently prepared to engage in the activities suggested. This can be decided only by well-established and trusted NGO workers and members of the village community. In villages where NGOs have not previously worked, outsiders cannot simply intervene with their own particular ideas and strategies without the risk of alienating local people who best know the characteristics of their lands.

There are many examples of NGO-facilitated community cooperative efforts to protect the environment in rural Thailand, which, although small in scale, can have significant local impact and taken together, can develop into a major social movement for environmental conservation.

DISCUSSION

Question: To what extent are women involved in the project?
Answer: Through participation in the various activities, women are taught hygiene,

nutrition, cooking techniques and are also given training in various income-producing crafts. Many of the key leaders and organizers of the project are women.

Question: Is Thailand's high environmental consciousness due to the country's laws?
Answer: Probably not as much as to the educational system; environmental education starts at the primary level and goes on through secondary school.

Question: Is Thai religion (i.e. Buddhism) a factor in the country's environmental awareness?
Answer: Religion may have some effect because of its stress on harmonious relations with nature.

Chapter Thirty-Six

Panama's ecological agenda: 1990–1994

RAUL H. PINEDO ANDERSON
Department of Environment and Natural Resources, Panama

PANAMA'S CURRENT ENVIRONMENTAL SITUATION

The destruction of forests is Panama's greatest environmental problem. In 1947, 70 per cent of the Republic was covered with forest. By 1987, only 43 per cent of its territory remained forested. Annually, this has meant a rate of destruction of 50 000 hectares, which indicates that by the year 2000, only 10 per cent of the country will have forests. The causes of this alarming deforestation include the spontaneous settlements that have grown out of extensive cattle ranching, as well as the widespread felling of trees for commercial timber.

One of the most dangerous consequences of deforestation is the loss of soils. Panama presents one of the most acute cases of soil erosion in Latin America. By 1960, 500 000 ha of the country's soil were degraded. Twenty years later, this area had expanded to 1.3 million ha, the majority in the Pacific area and the so-called "interior" of the country. By the year 2000, it is certain that 30 per cent of Panama's territory, about 2.0 million ha, will be seriously affected by erosion and incapable of supporting agricultural or cattle-ranching activities.

Panama depends increasingly on imported forest products. In 1960, the importation of wood products and sub-products cost US$ 4 million. In 1980, the country had to buy US$ 61 million worth of such goods. By the year 2000, forest imports will cost US$ 100 million.

Deforestation is also causing the extinction of the country's rich flora and fauna species. Panama has a great variety of natural habitats; it has the greatest coastline mileage, number of islands and mangrove forests in the whole of Central America. Scientists have identified 30 000 plant species, 225 mammalian species and 840 bird species. By 1980, hundreds of species of fauna alone were threatened. Between 1950 and 1980, the country's mangrove forests had been reduced from 500 000 to 200 000 ha.

ECOLOGICAL AGENDA

To ensure the protection of Panama's rich ecological legacy, freshwater and forest supply, we are working on the following agenda:

A national system of protected wild areas

Panama has dedicated 1.2 million ha of its territory to parks and wildlife reserves. The three most important national parks are:

- Darien National Park, on Panama's frontier with Colombia, has been declared a World Heritage Site and Biosphere Reserve. This park, *inter alia*, confines hoof and mouth disease. If this disease were to reach northwards to attack the cattle of the United States, there would be a direct loss of $3 billion and an indirect loss of $7 billion spent on activities to combat it during its first year.
- La Amistad International Park, on the Costa Rican frontier, recently declared a World Heritage Site. It contains water reserves that constitute hydroelectric power of the country.
- The Chagress National Park, whose forest protects the water source for the Panama Canal. Consequently, from an economic point of view, this is the country's most important park.

The National System of Protected Areas will permit the development of ecological tourism, which could generate an annual income of US$100 million by the year 2000.

Forest plantations and natural forest reserves

To guarantee the wood necessary for industrial use, by 1995 Panama must plant 50 000 ha of trees for sawmill and other uses.

The Cativo (*Prioria copasifera*) forest supplies 95 per cent of the resources for the plywood industry. Irrational use has caused a diminution from 70 000 ha in 1970 to 30 000 ha in 1990. Well managed, the Cativo could supply the national plywood industry on a sustainable basis.

Watershed management

Panama is subdivided into 51 watersheds, 18 on the Atlantic side and 33 on the Pacific side. The Department of Environment and Natural Resources has pinpointed six for special measures because of severe water conflicts, soil erosion and deforestation: the Bayano, the Panama Canal, the Rio Grande, La Villa and the Chiriqui Viejo.

We will strengthen the inventory and monitoring of drinking water quantity and quality in the watersheds. This will benefit 1.2 million people, half the national population. Soil conservation measures will concentrate in the Pacific area in order to minimize the expansion of the existing 1.2 million ha of degraded soil.

To achieve all this, the government of Panama is considering entering into a debt-for-nature swap arrangement with the United States, even though a country whose currency is soft is always disadvantaged in dealing in US dollars. The governments of Costa Rica, Ecuador and Uruguay, having done so, feel that they are paying more than necessary. This, among other reasons, accounts for Panama's reluctance to decide upon such an arrangement. It may well be a necessity. If so, the government hopes that its investments will be offset by other donor aid.

DISCUSSION

Question: Why does Panama consider entering a debt-for-nature swap when its government knows that it would suffer financially?

Answer: Costa Rica, Ecuador and Uruguay have had that experience. Panama is waiting to do so; dealing in US dollars always puts the developing country at a disadvantage.

Question: What has been done to protect the mangrove forests?

Answer: The International Tropical Timber Organization made a grant to take a mangrove inventory, a prerequisite for proper mangrove management.

Chapter Thirty-Seven

Regional approaches to planning: the African Ministerial Conference on the Environment (AMCEN)

HASSAN GUDAL
Formerly, Associate Director and Secretary of AMCEN
Regional Office for Africa, UNEP

Africa has been described as a continent in crisis. In 1990, the Executive Director of UNEP, Dr M. K. Tolba, stated that "it is suffering from a continual drain on, and degradation of its natural resources—plant cover, soils, water, animal resources and climate". Over-cropping, overgrazing, deforestation and general soil degradation destroy the resource base and cause the arable land to succumb to the encroaching desert. Each year, some 6 million ha of formerly productive land are reduced to sand and a further 21 million are reduced to a condition of zero productivity. A recent study shows that the desert is encroaching on other lands in Africa at the rate of 8–10 km per annum. Tropical forests are being lost at the rate of 1.3 million ha per year. The loss of soil cover brings with it a host of other problems such as the loss of soil fertility, soil erosion and the loss of genetic resources as the natural habitat is destroyed. Arid lands are rapidly spreading on a large scale. In the Sahel zone, south of the Sahara, we are witnessing a shift of ecological zones—desertification of the Sahel, sahelization of the savannah, savannization of the forests. In no other continent is the growth of the desert so severe. It is obvious that as a result of these factors, coupled with mismanagement and lack of adequate national policies, including the incorporation of environmental concerns in national development plans, Africa's once rich natural resource base and natural ecosystems have been overtaxed and overexploited to the point where valuable life support systems built up over many centuries have now been seriously damaged. The degradation of land in Africa is one of the most alarming features of the African crisis. This general degradation has led to the poverty of the African people and to lowering their quality of life. The continent of Africa has the largest number of the poorest countries in the world. There are 42 nations in the world known as the Least Developed Countries (LDCs), the poorest among the poor; of these Africa has 29. By comparison, Asia and the Pacific have 12 LDCs and Latin America only one.

To make things worse, the African continent has been receiving less rainfall since 1968. The flow rate of rivers and the level of lakes have dropped, adversely affecting fish resources, farming using controlled flooding, food and fodder production, river transport, energy production and groundwater recharge. The continual land degradation that was exacerbated by the decreasing rainfall culminated in the severe drought of 1983–1984, which brought about starvation and human suffering on an unprecedented scale. The drought need not have led inevitably to famine if the region's natural resources had been properly managed.

However, there is no doubt that Africans can reverse this destructive trend by taking the necessary steps to conserve their croplands, grasslands, forests, watershed systems and fisheries. This reversal will, of course, take time and enormous effort, but concerted action can no longer be delayed if environmental disaster is to be avoided.

Taking immediate action against this environmental degradation is exactly what the African governments did in 1983 at the peak of the 1983–1984 drought and famine when they requested UNEP to help Africa develop, formulate and implement a regional programme of cooperation to deal with the persistent and worsening common environmental problems in the region. UNEP swiftly responded to this request by initiating the necessary sub-regional and regional meetings and investigations which culminated in the Conference of African Ministers on the Environment, known as AMCEN, held in Cairo in December 1985. This ministerial conference evolved and adopted the Cairo Programme for African Cooperation to address environmental problems on the continent. The secretariat coordinating the implementation of this Programme is being run by UNEP until it will be taken over by the African governments some time in the near future.

The Cairo Programme has been developed to mobilize African resources and skills for sustainable development at all levels. It is intended as a tool to control and reverse the prevailing environmental degradation and the concomitant cyclical droughts and famine bedevilling the region. Its objective is therefore to halt this environmental degradation in order to rehabilitate the land and enhance its capacity as steps towards achieving self-sufficiency in food and energy.

The Programme is being implemented by six structural and three operational organs. The structural organs aim to mobilize available national and regional capabilities to implement and follow-up the Programme effectively. These structural organs include:

1. The Ministerial Conference, which is the policy-making body. It meets once every two years to review the Programme and decide on future action.

2. The Bureau, which is responsible for ensuring the implementation of the decisions of the Conference. Meeting once a year, the Bureau is composed of six ministers selected on a sub-regional basis;

3. The Secretariat, which coordinates the day-to-day implementation of the Programme;

4. The African Technical Regional Environment Group (ATREG), which is made up of African experts who advise AMCEN and assist in identifying problems and formulating proposals for adoption by the Conference;

5. The UNEP Task Force, composed of UNEP technical staff who give the necessary technical back-stopping to the AMCEN Secretariat;

6. The Inter-Agency Working Group (IAWG), made up of representatives of the United Nations' agencies and other international organizations who are responsible for identifying the role of their respective organizations in the implementation of various components of the Programmes as well as providing scientific and technical advice. This Group meets annually.

The operational organs consist of five Regional Committees, eight Regional Networks, 150 pilot village projects and 30 pastoral zone projects.

The Committees which are chaired by members of the Bureau have been established to concentrate on five major African ecosystems, namely:

- Deserts and arid lands;
- River and lake basins;
- Forests and woodlands;
- Seas; and
- African island ecosystems.

The Committees have developed workplans and sub-regional priority activities geared to the functions expected in their specific areas. For example, the Committee on Deserts and Arid Lands has prepared three sub-regional action programmes:

- The programme of conservation against savannization and sahelization in the central African sub-region;
- The outline of an Action Programme to combat desertification and promote food production in the southern African sub-region; and
- A draft Master Plan for development of the Nubian Aquifer for combating desertification in north-east Africa.

The Committee on River and Lake Basins has participated in the preparation of the Lake Chad Basin Master Plan and the Zambezi River System Action Plan. The Committee on Forests and Woodlands is preparing national reforestation plans for Nigeria and Zambia.

In order to develop and strengthen technical cooperation among African countries, the following eight Regional Technical Cooperation Networks were created to strengthen cooperation among scientific and technical African Institutions by exchanging information and conducting the basic studies and scientific research necessary for the environmentally sound utilization of resources:

1. Climatology;

2. Environmental monitoring;

3. Soils and fertilizers;

4. Water resources;

5. Environmental training;

6. Energy;

7. Science and technology; and

8. Genetic resources.

Each Regional Network is established in a suitable national institution that serves as a Regional Coordination Unit (RCU). Other similar national institutions in each country become members of the Regional Network. Most of the Regional Networks are hosted by existing national institutions, while others are administered for the time being by United Nations' agencies until such time as they are taken over by suitable national institutions (at present these Regional Networks are in different stages of functionality with outputs from them ranging from considerable to nil).

The aim of the Pilot Village Programme, which covers 150 villages and 30 pastoral areas, is to achieve self-sufficiency in food and energy in these villages and pastoral communities by using the traditional skills and experience of the villagers and pastoral peoples themselves in an economically feasible, environmentally sound and socially acceptable manner. These 180 national pilot projects have been launched as model activities to be replicated in other villages and pastoral zones in each country.

The ultimate goal of the AMCEN Pilot projects is the development of vigorous, productive communities that will provide their members, especially their young people, with opportunities to live well on the land rather than migrate to the cities. The 150 villages will represent the broad range of African ecosystems: arid lands, savannah, forests and woodlands, and coastal areas, among others. Similarly, the 30 pastoral areas will cover the full spectrum of rangeland–livestock environments in Africa.

In 1990, 29 pilot projects were being implemented in nine countries: Djibouti, Egypt, Ghana, Kenya, Senegal, Sudan, Uganda, Zaire and Zimbabwe. Representatives of these villages visited China in 1990 to gain from the experience of the Chinese, who successfully established similar villages that became self-sufficient in food and energy.

Even though the implementation of the Cairo Programme for African Cooperation is well under way, action is needed and could be taken at several levels to help facilitate the transition to sustainable development. First and foremost, the African governments themselves and also the international agencies and institutions and donors should undertake concrete measures to support the Cairo Programme, by, for example:

1. Providing support to AMCEN activities which involve direct grassroots assistance in different villages and communities to assist with the achievement of self-sufficiency in food and energy;

2. Strengthening the AMCEN Regional Technical Cooperation Networks, which address African priorities through research and training;

3. Assisting in expanding African programmes requiring sub-regional actions on Africa's major ecosystems such as the integrated development of river and lake basins; regional seas programmes; deserts and arid lands; forests and woodlands; island ecosystems.

The support should include financial, manpower, technical and advisory inputs. Africa's economic development rests on the salvation of its environment.

Chapter Thirty-Eight

The role of law in environmental management at the national level

IWONA RUMMEL-BULSKA
Coordinator, Interim Secretariat for the Basel Convention, UNEP

INTRODUCTION

Only in the early 1970s did the world begin to speak of environmental law as such, as a new and separate branch of jurisprudence. However, environmental laws have existed for millenniums. In ancient Egypt, decrees prohibited disposing of many toxic wastes in the Nile. Sri Lanka kings in the fifth century set aside huge tracts of land as wildlife sanctuaries and guarded them rigorously. In 12th-century London, the burning of coal was strictly regulated because it was recognized as an air pollutant. Largely through custom over centuries, European peasants fallowed strips of land on a rotating basis to prevent soil exhaustion. Beyond written ordinances and customary norms lay religious practices; dietary strictures, whatever their theological significance, controlled food and thereby protected health. By and large, throughout history world-wide until fairly recent times, lawyers were simply people who wrote down existing rules for behaviour under certain conditions. They did not seek to modify behaviour or foster its change—a point to which we shall return.

Only in the mid-1960s did a true conception of the environment itself arise, probably because industrial development in the north-western world accelerated so rapidly after 1945 that citizens of that region perceived the degradation of their physical surroundings on a palpable, colossal scale. Consequently, some began to speak of bringing together their countries' bits of law relating to the environment. As a mosaic of these disparate regulations emerged in several nations, it was evident that many affected others, that there were gaps, overlappings and even contradictions among them. It was recognized, too, that if these bits of legislation were not synthesized within a comprehensive approach to the environment, serious problems could arise, some of them irreversible.

Strange as it may seem, the Stockholm Declaration on the Human Environment of 1972 contains no explicit reference to international environmental law, largely because the vast majority of international legal experts at that time could not accept this concept. They regarded the many statutes related to the environment as belonging to juridical domains such as water law, labour legislation, land tenure and usage and urban regulation. Nevertheless, several countries wrote into their constitutions the right of their citizens to a decent environment or physical conditions. Now more than 30 nations have such constitutional provisions, while over 100 refer to the right to health or decent conditions of life.

"HARD" AND "SOFT" LAW

Law tends to fall into two categories, "hard" and "soft". The former is legally binding and contains measures for its enforcement. The latter aims at moral commitment, changes in attitude and behaviour and usually takes the form of guidelines or principles. For example, air quality law tends to be "hard", while legal action on noise, a relatively new environmental concern, may well be treated in terms of guidelines. Although the relative merits of "hard" and "soft" law are still hotly debated, the latter often leads to the former. The choice of one type rather than the other generally stems from the power of various lobbies and the weight of public opinion on any given issue. Non-governmental organizations play a significant role in shaping public opinion, even though their views are sometimes unrealistic. Indeed, UNEP has occasionally called NGOs and industry together to discuss issues for which either "hard" or "soft" law was required and urges governments to do likewise.

DEVELOPMENT OF NATIONAL ENVIRONMENTAL LAW AND INSTITUTIONAL STRUCTURES

During the 1970s, most countries began assembling existing laws related to the environment—notably legislation concerning health and sanitation—and revising them within a comprehensive framework. Others decided that their legislation was so outdated that they formed groups of lawyers to develop bases for national environmental law. Still others asked UNEP to guide their elaboration of umbrella legislation. For instance, UNEP assisted Uganda to combine its customary and colonial laws, all of which were reviewed and revised to produce a National Environmental Act.

A single comprehensive national environmental act is very useful and convenient, but also very difficult to attain, largely because few governments have the institutional structures through which such legislation can be smoothly developed and properly enforced. Although many governments have established environmental ministries, far more have provided for special environmental bodies within the more traditional ministries such as those of natural resources, planning or industry. The reason for this is usually economic. The environment is not perceived as generating income and therefore revenues. However, as environmental concerns now penetrate virtually every sector, existing ministries have broadened their respective mandates to encompass these concerns. In many cases, such horizontal expansion has led to the creation of a special inter-ministerial commission or committee, often close to the head of state or government.

Given such an institutional structure or a comparable organization of governmental bodies, the evolution of national environmental law may well begin with a *set of standards* from which specific sectoral legislation can derive. In this case, it is always useful to think first and foremost about protecting the widest possible public. The labour force may be a good point of departure, in part because workers can and should organize to support legislation on the occupational environment.

At present, a number of countries are formulating *national strategies* to ensure the integration of environment and development. In several instances, this new approach has grown out of the relatively new concern about the destruction of biological diversity, but it has other broader sources as well. Each nation must base its approach on what best suits its needs and capacities, including its capability to monitor such vital aspects of its environment as air and water quality.

Similarly, enforcement measures will vary from country to country. A few basic forms are taxes, price controls and licensing systems, as well as civil liabilities and criminal penalties. Each government will also have to choose what kinds and degrees of monitoring and control it wishes to delegate to the country's provincial and local authorities. In this connection, too, it is useful to think in terms of the widest possible outreach to develop appropriate enforcement. Women at the grass roots level, for example, can often prepare effective means of monitoring local environmental trends, as well as penalizing environmental offenders for the damage and destruction wrought, simply because they manage households and, increasingly, agricultural land.

Thus, law can perform a variety of functions, ranging from the protection of the environment to creating awareness of its value for a broad spectrum of citizens— in short, fostering the ethic of stewardship our battered planet so badly needs.

Further suggestions may be found in two UNEP publications *New Directions in Environmental Legislation, Particularly in Developing Countries* and *Legal and Institutional Arrangements for Environmental Protection for Sustainable Development in Developing Countries*.

DISCUSSION

Question: Even though a country has adopted sound laws and regulations, their enforcement is hampered by major problems in the legal system itself. Can UNEP help with such difficulties?

Answer: Both UNEP's Environmental Training Unit and its Environmental Law and Institutions Programme Activity Centre can assist governments upon request. Legal training and sensitization can be undertaken not only for lawyers but for judges as well. Certain countries may wish to select one of their own law schools and there establish a comprehensive programme of training on different aspects of the environment. Inspectors, too, must be trained, but this task is far more technical because it is so closely related to specific aspects of environmental conditions that prevail within the country; it would therefore be more effective if undertaken within the country itself. In addition, countries may well have to adopt additional measures concerning non-compliance of industry and governments themselves with national and international environmental laws.

Question: How can training for both legal and monitoring functions be financed?
Answer: Training, as well as the equipment necessary for proper monitoring—notably the latter—may be purchased or supported by private enterprise. Industry itself could be asked to purchase the equipment and prove to government authorities that they are complying with the law. Another approach would involve contributions by industries to buying monitoring equipment for government use.
Comment: A training levy can also be imposed on products or services provided by certain industries. In Kenya, for example, a training tax is added to restaurant bills, so that catering becomes a source of revenue.

Question: Should standards be uniform throughout a country or should provinces be allowed to establish their own standards?
Answer: It depends upon the size of the country. Different categories of standards may well be established for different components of the environment. For example, with regard to water, very strict standards might be imposed for such areas as protected lands and recreational zones, more flexible ones for other parts of the country or various uses of water. However, the stringency of the standards should stem from the feasibility of their implementation. Unrealistic standards eventually create disregard for the law.

Chapter Thirty-Nine

Designing, developing and obtaining international legal instruments for environmental protection

MOSTAFA K. TOLBA
Former Executive Director, UNEP

INTRODUCTION

Let us not enter into any discussion of the merits of the various types of normative activities needed for international environmental protection, notably comparisons of "hard" and "soft" law. The evolution of international environmental jurisprudence—any agreement involving two or more countries—defies traditional definitions. What defines the need for any type of instrument is the nature of a particular problem and its magnitude. Most important, no treaty in and of itself ever solves a problem.

REGIONAL AGREEMENTS

UNEP's first ventures into international law concerned regional seas, initially the Mediterranean, whose very name indicates that it is enclosed by land and consequently is a body of water with obvious geographic boundaries. From 1968 into the early 1970s, increasing outcries rose from scientists and others concerned with activities ranging from fishing through recreation that the Mediterranean was dying, choking to death on pollution. The problem was aggravated by the fact that this particular sea is so enclosed that its waters change only every hundred years. However, its very enclosure by 18 nations—all of which had, at one or another point in history, strong cultural and commercial ties, whatever their relations by the mid-twentieth century—gave rise to the concept of the Mediterranean as a common heritage.

In 1974, UNEP began examining the Mediterranean's sources of pollution. Three were obvious: land-based wastes such as industrial and untreated sewage; agricultural run-off, particularly pesticides and fertilizers, from the region's many rivers such as the Nile; and spillage of many types, notably oil, from ships. These three kinds of pollution alone constituted the equivalent of a *force majeure* with which

UNEP confronted the governments concerned. These included quite a number of belligerants, among them Israel and certain Arab lands; Turkey and Greece in conflict over Cyprus; and Morocco, Tunisia and Algeria, engaged in territorial, trade and other rivalries, as well as problems with France. Indeed, virtually the only countries whose relations seemed harmonious at the time were Italy and Spain. None the less, the fact that all these countries came to see the significance of working together to save a shared natural legacy essential to them all became the first strong signal that *the environment could be a unifying rather than a divisive factor.*

The Convention for the Protection of the Mediterranean Sea against Pollution, adopted in 1976 in Barcelona, Spain, was the first *binding* accord through which the 18 governments of that seaboard agreed to share the work required to save their common source of life, the richer countries of the region providing much of the financing and technical assistance to the poorer ones, all through a special United Nations' trust fund, within the framework of the UNEP-administered Mediterranean Action Plan. Moreover, the Barcelona Convention became the umbrella for a number of additional binding agreements—protocols concerning problems that range from dumping from ships to specially protected areas.

Very shortly thereafter, UNEP was asked to help formulate a comparable plan by the eight nations surrounding that all-but-enclosed sea whose very name is a source of contention. United Nations terminology designated it the 'Persian Gulf', but the Arab countries on its southern side refused to speak of it as such. The name they preferred, the Arabian Gulf, was equally objectionable to Iran. And when UNEP tried to resolve this disagreement by talking about the Persian–Arabian or Arabian–Persian Gulf, both sides balked.

Indeed, the nomenclature of the Gulf held up negotiations on an action plan for some two years, despite the fact that the initial assessments of its water quality— undertaken several years before this was further damaged by the Iran–Iraq war— showed that it was 40 times more polluted than any body of water on Earth of comparable size. This alarming condition stemmed not only from its shape—long and narrow—but, far more important, its use as a passageway for huge oil tankers that took advantage of the border countries' lack of proper coast guard services. Tanker captains even *washed* their vessels in the Gulf. Moreover, industrial development in the area was accelerating rapidly.

When agreement on an action plan was reached at last in 1978, it was named the Kuwait Regional Convention for Cooperation on the Marine Environment, simply because the instrument was signed in Kuwait. Similarly, work is coordinated by the Regional Organization for the Protection of the Marine Environment (ROPME), located in Kuwait and financed by the parties through another special United Nations' trust fund.

Regional seas conventions now involve more than 120 countries and cover the wider Caribbean, the Atlantic coast of west and central Africa, the Pacific Coast of South America, the islands of the South Pacific, the East Asian region, the South Asian seas, the Red Sea and the East African/Indian Ocean islands' region. In a parallel effort for shared *freshwater resources*, UNEP launched its programme for the environmentally sound management of inland waters (EMINWA) in 1984. The first agreement of this nature brought together the eight nations of the Zambezi River basin in 1987.

GLOBAL PROBLEMS AND TREATIES

A similar but far more complex pattern emerges when one turns to environmental law on global issues. These often involve North–South disputes, best approached initially by "soft law", i.e. international guidelines, principles and standards.

Two salient problems are the trade in toxic substances—chemicals manufactured in developed nations whose sale is often banned in their respective countries of origin—and the transboundary dumping of hazardous wastes, a crime that received immense media coverage during the 1980s because it became as dangerous as the drug and arms trades. It often involved buying tracts of dumping space in very poor countries without their governments' knowledge by one-desk private companies for the movement of wastes whose disposal was not permitted in the countries of origin. Moreover, these illicit operations were carried out at a mere fraction of the cost of safe disposal.

"Soft" law is negotiated by government representatives just as if it were "hard"— which, indeed, it often becomes within several years. In the case of dangerous chemicals, the UNEP International Register of Potentially Toxic Chemicals (IRPTC) brought expert groups together to elaborate the Provisional Notification Scheme for Banned and Severely Restricted Chemicals, adopted by the UNEP Governing Council in 1984 and followed three years later by the London Guidelines for the Exchange of Information on Chemicals in International Trade. Both made arrangements for information sharing on legal controls and exports and imports of dangerous chemicals between national authorities. Both were carried out in close cooperation with a variety of international bodies, notably FAO, ILO, WHO, and OECD. However, a sensitive issue, particularly for developed countries, was the principle of *prior informed consent*—a detailed explanation of the negative impacts of a given product, complemented by a methodology for dealing with them, on the basis of which the importing country can decide whether or not to buy the substance. Two years of tense and intense persuasion were needed to obtain agreement on this type of procedure.

Similarly, action on the dumping of wastes, which culminated in the 1989 Basel Convention for the Control of Transboundary Movements of Hazardous Wastes and their Disposal, began with issuing guidelines on the subject, in this case the 1987 Cairo Guidelines and Principles for the Environmentally Sound Management of Hazardous Wastes. The Convention, of course, features the concept of prior informed consent, which is now being applied in virtually every subject area to which it is pertinent.

As indicated above, the resolution of issues that involve disagreements between North and South usually begins with "soft" law. However, when a problem is clearly shown by science to involve world-wide suffering, "hard" law—treaty—negotiations generally begin immediately. The dangers to human health, as well as that of whole ecosystems, caused by the depletion of the ozone layer is perhaps the most obvious of such issues. None the less, the formulation of the 1985 Vienna Convention for the Protection of the Ozone Layer, *which covered only research and exchange of information*, required two years of intensive negotiation. UNEP immediately set in motion the elaboration of a protocol that specified *procedures and targets for the elimination of ozone-depleting substances*, chlorofluorocarbons (CFCs) and halons. Though at the time, world-wide production of these chemicals involved only some US $2 billion, their use in a variety of manufactures—notably coolants, microchips and large-scale fire-extinguishing agents—multiplied this sum several hundredfold. Consequently, the elimination of these chemicals posed an enormous threat to industry, employment and trade.

For two years, one after another, economic and social problems emerged from every regional group. Their numbers and natures were dwarfed by the magnitude of the threats to human health alone that science had clearly revealed. But if one thinks of a working parent in any country, rather than some huge transnational conglomerate, that parent will probably choose his or her job above what appears to be a remote health impairment. Bearing that in mind, UNEP pursued constant informed negotiations that eventually built up the confidence in its representatives leading at last to minimal agreement in 1987: the Montreal Protocol on Substances that Deplete the Ozone Layer. It was the world's first pre-emptive environmental treaty. Moreover, reassessment and revisional provisions were contained in the instrument itself.

By 1989, with the discovery of the Arctic as well as Antarctic ozone hole, northern NGOs pushed their governments further, so that by 1990, the built-in reassessment produced the London amendments to the Protocol. These called for a complete phase-out of 87 chemicals by the year 2000, added many other substances to those originally listed and established a Trust Fund that allows developing countries to buy substitutes without adding to their economic burden.

Further scientific assessment completed in late 1991 revealed far more alarming discoveries. It poses new questions as well. Will an earlier phase-out date for today's controlled substances speed the repair of the ozone layer? Will adequate substitutes be available in time at a reasonable cost? If both answers are positive, a still stronger Protocol may well emerge. And from that strengthened or otherwise changed instrument, still other questions will emerge. *Environmental law must continuously evolve in keeping with the findings of science.*

Let is never be forgotten, too, that the strength of any treaty lies not only in its provisions but in the number of its parties and the vigour of their implementing those provisions. As yet, the effectiveness of the Montreal Protocol relies very heavily on the *good faith* of its States' Parties. At present, its compliance procedures are quite weak. Governments that have accepted the on-site verification procedures of disarmament treaties are still reluctant to accept that principle in laws concerning the environment. If they do not do so—and soon—we may well face world-wide catastrophe as grave as a nuclear holocaust.

DISCUSSION

Question: Is there any conflict between the Basel and Bamako conventions on hazardous wastes?
Answer: None. The latter is simply more stringent, an eventuality provided for by the Basel treaty. In addition, it contains an element not covered by the Basel Convention: nuclear wastes, which are dealt with by another international body, the International Atomic Energy Agency.

Question: How can the non-compliance provisions of the Montreal Protocol be strengthened?
Answer: The Protocol itself envisages an Implementation Committee to verify compliance. In addition, further agreements can be included for on-site verification.

Question: Has UNEP developed any strategy to convince countries that have not so far ratified environmental treaties to do so?
Answer: We are now concentrating on securing enough ratifications for these conventions' entry into force. Once that has happened, we can discuss in depth appropriate incentives or other actions to induce strict adherence to these instruments.

Question: Can environmental conventions be used as a form of political pressure on developing countries in terms of their receiving foreign aid?
Answer: That is most unlikely. The Trust Fund created by the amended Montreal Protocol calls for a committee composed equally of developed and developing countries to decide on the allocation and use of moneys. All decisions must be made by two-thirds of its membership.

Chapter Forty

The Global Environmental Facility (GEF)

MIKKO PYHALA
Chief, Clearing-House Unit, UNEP

Initially proposed by the government of France in 1989 and rapidly seconded by that of Germany, the Global Environmental Facility (GEF) was a three-year experiment established in 1990, administered by the World Bank, UNDP and UNEP, providing grants for investment projects, technical assistance and research in developing countries in the following four environmental problem areas:

- global warming, particularly the effects of greenhouse gas emission on the world's climate and the destruction of carbon-absorbing forests;
- the pollution of international waters through oil spills and the general accumulation of wastes in the oceans and international river systems;
- the destruction of biological diversity through the degradation of natural habitats and the "mining" of natural resources;
- the depletion of the ozone layer from emissions of chlorofluorocarbons (CFCs), halons and other gases.

The Facility had at its disposal US $1.3 billion in Special Drawing Rights (SDRs) for its initial three-year pilot phase. Essentially, the GEF is an umbrella made up of resources from the following three sources:

- a core fund known as the Global Environmental Trust Fund (GET), which comprises US $800 million in commitments;
- several co-financing arrangements that can provide approximately US $300 million on grant or highly concessional terms;
- some US $200 million provided under the Montreal Protocol to help developing countries comply with its provisions to phase out substances that destroy the ozone layer. These particular funds are administered by UNEP under the auspices of a 14-country Executive Committee.

Although no set formula exists for the allocation of these resources, 40–50 per cent were foreseen for efforts to reduce global warming, 30–40 per cent for the conservation of biological diversity, and 10–20 per cent for the protection of international waters. These outlays were not perceived as altogether satisfactory to many developing countries, which would have preferred to have seen provision made for land degradation as well, notably deforestation and desertification. However, a number of projects that concentrate on these areas of concern could, theoretically, qualify for funding under climate protection and activities to preserve biological diversity.

As indicated above, responsibility for administering the GEF was shared among UNDP, UNEP and the World Bank on the understanding that the operation of the Facility would create no new bureaucracy in and of itself and that the three implementing agencies would make only modest organizational changes in their own structures to fulfil their mandates in respect of the GEF. Generally, these mandates were defined as the following:

- UNDP would be responsible for technical assistance activities and, through its world-wide network of offices, help to identify projects through pre-investment studies. It was also charged with running the small grants programmes for NGOs, which will be discussed further below;
- UNEP provided the secretariat for the Scientific and Technical Advisory Panel (STAP), as well as general environmental expertise. In this connection, UNEP hoped to steer GEF investment towards innovation;
- The World Bank administered the Facility, acted as the repository of the Trust Fund, and was responsible for investment projects.

This was not an exclusive arrangement, GEF projects may be sponsored by regional development banks and by other specialized United Nations bodies, including those responsible for food, agriculture, health, climate and maritime activities.

Moreover, from the time the GEF was launched, the three prime implementing agencies committed themselves to working with NGOs whose specialized expertise could help significantly in identifying, reviewing, preparing and implementing projects, the last particularly in biodiversity. The implementation of all projects financed through the $5 million small-grants programme was the responsibility of the NGO beneficiaries. This pilot programme, scheduled to operate initially in 35 countries, consisted of individual grants of up to $50,000 for community-based activities and regional and sub-regional projects of up to $250,000. In much larger projects, too, it must be pointed out, popular participation is of the utmost importance.

Eligibility criteria were both simple and complex. All countries with a per capita income of less than US $400 annually (as of October 1989), as well as a UNDP programme in place, qualified for GEF funds, as did projects that were deemed to benefit the global environment, as distinct from the local environment, provided that they fell into the four priority areas outlined above. However, not all projects that benefit the global environment automatically qualified for GEF support; they also needed to be *innovative* and demonstrate the effectiveness of a particular technology or approach. These criteria also included a project's contribution to *human development* through education or training, as well as its provision for evaluation and demonstration of results.

Further, projects that were viable on the basis of local costs and benefits were not necessarily eligible for GEF funding, whatever their benefits to the global environment. Conversely, GEF resources were provided if a project offered substantial global benefits, but was unlikely to be viable without some concessional funding. Similarly, a project that was economically viable, but required supplementary funding to bring about global benefits qualified for GEF assistance.

At least initially, the cost-effectiveness of GEF investments was determined on the basis of physical rather than monetary yardsticks of global benefits. Consequently, for a climate-related project, the major measure was the reduction in carbon emissions, while a project concerning marine pollution was evaluated in terms of the amount of ship-generated waste disposed of or the expected improvement in the health of a given ecosystem. None the less, an excessive price was not to be paid to achieve a given physical benefit.

Finding out what works—and why—was crucial to the GEF process. Information obtained in the monitoring, review, supervision and evaluation of projects will be used to refine GEF guidelines and procedures, to evaluate different technologies and to determine where the Facility's intervention has been most successful.

By mid-1994, the end of the GEF pilot phase, all funds were to have been committed, even though actual disbursements will probably continue through 1997 or 1998. The lessons learned will constitute a base for cooperation in the years to come. Central to those lessons is the challenge to create the policy frame-work and institutional capacity that furnish a creative mix of incentives and disincentives, of regulations and market mechanisms to give future projects the opportunity to provide viable, long-term solutions to our fundamental environmental problems. To give but one example, new techniques for generating energy, such as geothermal generators and photovoltaic cells, will not displace fossil fuels if national pricing policies provide no incentive to produce environmentally sound alternatives. The GEF should be regarded as a testing-ground to persuade people of the importance and practicality of long-term environmental management.

DISCUSSION

Question: GEF criteria are often incongruent with the problems and priorities many developing countries must face. Will these criteria be revised to respond to those needs?

Answer: In many instances these criteria are not quite as rigid as they may seem at first glance, although many discrepancies may continue to arise because the donors wish to focus only on global issues. However, the donors increasingly realize that these issues cannot be tackled unless the economic and social problems of individual developing countries are addressed. None the less, the GEF will continue to be concerned with global environmental problems and must not be regarded as a funding system for development.

Question: Do proposed projects for GEF funding have a better chance of approval if they are linked to other World Bank operations?
Answer: Yes. The Facility can fund projects up to $10 million if they are free-standing, but up to $30 million if they are linked to a World Bank operation.

Chapter Forty-One

China's population, environment dynamics and the impact of climate change

XIA GUANG
National Environmental Protection Agency, China

INTRODUCTION

Throughout history, China has been a populous country. Its population was 540 million in 1949 when the People's Republic of China was established. By July 1990, the population had reached 1.16 billion, two significant spurts of population growth having occurred during the 1950s and 1960s, respectively. China now accounts for about 20 per cent of total world population. China's land area including its inland waters is approximately 9.60 million sq. km—in short, the third largest territory of the world. This land is enormously diverse in its topography. It includes mountainous regions (33 per cent), hills (10 per cent) plateaux (26 per cent), basins (19 per cent), plains (12 per cent) and a coastline of 18.4 thousand km.

Because of its environmental diversity, both natural and anthropogenic, the relations of the Chinese population to its settings have been distinctive and vary from one region to another. For example, on the alluvial plain of the Yangtze and Zhujiang Rivers in the southeast plain of the country, the dense agricultural population and intensively cultivated paddy fields and low hilly lands constitute a fairly potentially stable relationship of population and environment. The sparse nomadic population and extensively cultivated dry and semi-arid grasslands of west China manifest a different but equally potentially stable relationship.

The population of China depends heavily on its natural environment, because it is fed basically by domestic rather than foreign resources, while Chinese agricultural yields vary significantly with weather and other fluctuations such as the frequent natural disasters. In addition, there is an obvious disequilibrium of population distribution in China, most of whose population is concentrated in the coastal regions. Ninety-one per cent of the population lives in the southeastern regions, which account for 43 per cent of the country's land. Consequently, there is a high environmental load on those regions that are also highly sensitive to external conditions. Further, population growth has exceeded the natural carrying capacity of

China's ecological resources. This overload exerts heavy pressures on different environmental elements such as land, water and atmosphere, making population/environment relationships extremely tense and fragile.

Solving the population/environment problem is a long-term endeavour. During the next two or three generations, further population increase will probably be inevitable, given the general inertia of population trends. Even where corrective measures are taken, many decades, a century, will pass before significant changes occur. The impacts of climate on the population of China and its environment reveal the following four trends.

DROUGHT

The north and northeast of China is rapidly becoming more arid, straining further the relations between population and environment. The impacts of climatic disasters on Chinese agriculture are serious. The farmland area affected averages about 33 million ha each year; this accounts for 36 per cent of the entire cultivated area of the country. The disaster areas (those that show a drop in production of at least one-third), amount to more than 13 million ha, which represent 14 per cent of all Chinese farming.

Seasonal drought caused by temperature increases would bring about many negative effects because the major natural disaster for agriculture in China is drought. There are frequent spring droughts in northern China and in the mid-north loess plateau, the southeast of Inner Mongolia and the western part of the northeast plain. According to predictions of climate change, the damage to crops caused by drought in these regions would increase by 5 per cent, wheat and spring-sown crops being most affected. The loss of output is estimated to be at least 4 million tonnes annually. The hot summer drought in July and August would be aggravated in the southern regions, the rainfall line shifting northwards deferring its return, thereby affecting the early rice maturation and the early growth of the late rice crop. It is estimated that output would be reduced by 20 per cent, which corresponds to a loss of 6 million tonnes of grain annually. Dry autumns would occur often in the western regions, affecting the growth of autumn crops. The seasonal southward return of the rainfall line would be postponed, aggravating the ravages and prolonging the period of drought. The projected global warming would also aggravate the harmful effects of the dry hot wind in the north. The grain output reduction is estimated at 5 per cent. At the same time, plagues of pests would increase and so would the cost for pest control.

To sum up, the comprehensive effects of climate change on Chinese agriculture would reduce the country's production potential by at least 5 per cent. Because of the population/land ratio in China, each hectare of cultivated land provides food for 11 people, a much higher level than the world average. The land loss caused by climate change would lead to radically lowered standards of living for much of the nation's rural population, some of whom might face starvation if they remained on their lands. In short, population/environment relations would face severe adjustments.

TYPHOONS

The arid zone of China is moving southwards, reducing the agricultural productivity of the alluvial belts of the Changjian River in southeast China, which is densely populated and now has a relatively high productivity. Population increase in these regions would have drastic effects.

The natural conditions in the southeastern regions of the country are highly favourable for human life. These regions are the most important areas for grain production in China, usually called "the land of fish and rice". Consequently, they are vital to the Chinese economy. In those regions, the prevailing agricultural patterns are those that produce water crops as food staples. Intensive cultivation has been going on for some two millennia. Population density, as indicated earlier, is high, as is relative production efficiency. There, a basically mutual adaptability and balance normally prevails between human beings and the environment.

In the southeastern regions that border the coastline, the normal summer temperatures are high. Global warming would lead to the formation of low pressure zones, resulting in an increase of frequency and intensity of typhoon attacks in these areas. According to the current estimates, 170 000 ha of farmland would suffer from these storms, and grain output would be reduced. This would doubtless alter population/environment relations even further, reducing the scope of human activities because even the current population growth rate has outstripped the land's carrying capacity.

SEA-LEVEL RISE

Moreover, the southeast coastal zone of China is the country's most developed area industrially and agriculturally. As indicated earlier, it represents 43 per cent of the nation's territory and is home to 91 per cent of China's population. The low coastal plain areas are found largely along the Bohai Bay and the estuary areas of the Yellow River, the Yangtze River and the Pearl River. Since ancient times, the Chinese have developed the coasts of the region into farmland aquaculture areas, fish being the important non-staple food of the coastal cities. Sea-level rise would submerge or severely change these coasts and the production they support. Their reconstruction would be very costly. Nearly half the area of the Pearl River delta (about 3500 sq. km) would be inundated. The more developed areas in the Yangtze River and Yellow River deltas could also be devastated. In addition, housing, transportation and the water supply for domestic, industrial and agricultural uses would suffer immensely. Salt water would invade the inland areas, increasing the range of salinized land, destroying coastal ecologies and those environmental protection facilities that exist. Hundreds of millions of Chinese would be forced to move inland, thereby creating yet another set of severe environmental stresses.

THAWING

China ranks third in the world in the extent of its tundra area. Global warming would thaw much of China's tundra and northern steppe, both east and west, which account for 18 per cent of the country's territory. A good part of this land would become desert. The livestock of its pastoral populations would perish.

If temperatures rise by 0.5°C and remain stable for 10–20 years, about 5 per cent of the tundra in China would thaw out. If temperatures rise by 2°C over a sustained period of 10–20 years, about 40–50 per cent of the tundra would melt, triggering enormous landslides and severely damaging housing, railroads and road networks. The thawing process would also lower the water table sharply. Grassy marshland on the plateau would become seriously degraded and vegetation in the high mountains would die. Finally, cold desertified areas would form, notably on the plateau, and pasture land would decrease radically.

In conclusion, the environmental picture in China contains an enormous population element. Consequently, China is paying great attention to any global changes, notably climate change, because of the drastic effects they may have on the country's peoples.

Chapter Forty-Two

Attaining sustainable development in Nigeria: water quality monitoring management

MODUPE TAIWO ODUBELA
Federal Environmental Protection Agency, Nigeria

Nigeria is one of the largest African countries. Its area totals 923 768 sq. km. It lies within the tropics along the Gulf of Guinea on the west coast of Africa, bordered on the west by the Republic of Benin, on the north by the Republic of Niger and the east by the Republic of Cameroon. The population of the country is about 100 million. Its major ethnic groups include the Hausa, Ibo, Yoruba, Fulani, Edo, Urhobo, Efik, Irjaw, Tiv and Kanuri.

Nigeria is naturally endowed with abundant surface and underground water resources. Two of the major rivers—the Niger and the Benue—flow through international boundaries. Lake Chad is also a large body of water strategically located and managed by Nigeria and her neighbours—Cameroon, Niger and Chad. Some of Nigeria's rivers are shared by its federated states, among these the Kaduna River, the Ogun River and the Imo River, as are many estuaries and lagoons along the 920-km coastline at the southern border of the country along the Atlantic Ocean. These water bodies are very rich in freshwater and aquatic resources.

WATER POLLUTION PROBLEMS

In developing countries such as Nigeria, the huge debt burden, population explosion, and moderate-to-rapid urbanization and industrialization trends have left our people too few options or none but to accept water sources of inferior and doubtful quality. Either alternative sources do not exist or economic and technological constraints prevent access to the available water that was adequate before the county had to face the immense problems mentioned above. In a FEPA report of 1991 Odubela *et al.* reported that institutional frameworks and regulations for water quality protection are lacking or, where available, poorly developed, uncoordinated and ineffective.

SOURCES OF WATER POLLUTION

Water is polluted in Nigeria by point and non-point sources. Surface waters are contaminated by human disposal of wastes from socio-economic and domestic activities. For example, Lagos alone disposes about 100 000 tonnes out of the 500 000 tonnes of refuse generated each year into the Lagos lagoon. Regrettably, all the streams and rivers in the major and coastal areas of Nigeria constantly receive the solid wastes that are complex mixtures of biodegradable organic materials, such as human faeces, dead animals, leaves, and abattoir wastes, as well as toxic chemicals, including pesticides and heavy metals.

High population pressure and rapid urbanization have resulted in the dumping of domestic and raw sewage into our lagoons, streams and rivers in the urban and rural areas. Some examples of the rivers in these urban areas are the Kaduna, the Niger, the Ogba, the Ogunpa and the Ogun. It is estimated that the Lagos lagoon at Iddor receives about 50 million litres of human faeces annually. This condition has arisen because of a lack of adequate sanitary facilities in the urban and rural settlements. Seepage from cesspools and soakaway pits and leachates from refuse dumps are major factors in underground water pollution in the heavily populated cities such as Lagos, Port Harcourt, Kaduna and Kano.

The continuous discharge of untreated industrial effluents laden with toxic chemicals, such as lead, mercury, cadmium, DDT and PCBs, into public drains and surface waters (rivers, streams) in these cities has rendered the rivers malodorous and murky, unwholesome for any beneficial human uses.

Petroleum exploration, exploitation and transportation activities with frequent oil spills, improper disposal of drilling wastes, and non-sustainable management of wastes from petroleum refineries and petro-chemical plants have also polluted the fresh and marine water sources of the coastal areas of Nigeria.

Non-point sources, such as run-off from farmlands treated with fertilizers and pesticides, have led to the eutrophication of surface waters and the infestation of freshwater rivers, streams and lagoons of the coastal areas of Nigeria, such as the Lagos lagoon, Badagry Creeks and rivers in Ondo, Edo, Ogun and Akwa-Ibom, with weeds, especially the notorious water hyacinth. Land degradation problems arising from bad land-use practices have resulted in soil erosion and siltation of rivers, fishery resources destruction, difficulty in transportation and impaired water quality.

ENVIRONMENTAL IMPLICATIONS OF WATER POLLUTION

The environmental implications are enormous both in public health and other socio-economic areas. According to Aina (1991), if water pollution in Nigeria is not checked now, it will cost the nation about US$ 10 billion per year. The losses from water hyacinth alone amount to about US $5 billion per year.

Less than 30 per cent of Nigeria's population in rural areas has access to safe drinking water because the nation's water supply has lagged behind its water needs.

Constrained by acute shortage of potable water for drinking, cooking, washing and for irrigation, people are forced to use the polluted water. Fortunately, Nigerians do not eat their food raw, so that the cooking process kills most of the pathogens in water. However, the old belief that dirty stream water cannot kill is no longer sustainable, particularly where drinking water is concerned. Accordingly, water-related diseases such as cholera, typhoid, hepatitis, dysentery, guineaworm and poliomyelitis are rampant. The World Bank has reported that about 40 million Nigerians are at risk from water contamination.

The loss of human resources due to premature deaths, the increased morbidity rate and the loss of work-hours due to absenteeism is enormously detrimental to the national goal of sustainable development. A great deal of the nation's wealth that could have been better used in other vital and productive sectors is being used to purchase drugs and vaccines for therapeutic and prophylactic purposes.

IMPORTANCE OF WATER-QUALITY MONITORING AND WATER-QUALITY DATA

Water quality refers to the extent and variety of foreign substances in the water. Water-quality monitoring is a vital aspect of water management because it is a tool for establishing baseline data for critical water pollutants and providing a basis for guiding the implementation of contingency and corrective plans against pollution. Nigeria has no organized national water-quality monitoring system nor a national water-quality data bank. The available water-quality data are scattered among various higher-education institutions, water corporations, river basin development authorities, the Ministries of Health and Water Resources and private firms. According to the GEMS/Water report, UNEP lamented the absence of Nigeria and some other African countries in global water-quality monitoring programmes.

NIGERIA'S STRATEGY FOR WATER-POLLUTION CONTROL AND WATER-QUALITY MANAGEMENT

Nigeria recognizes the intrinsic linkage between water quality and water quantity and the need to preserve the integrity of the aquatic ecosystem for sustainable development. Therefore, in 1988, the Nigerian Government established the Federal Environmental Protection Agency (FEPA), thus providing the effectively coordinated institutional and legal framework for environmental degradation and pollution control that was lacking. The Agency has the overall responsibility for environmental protection and environmental technology matters in the country. It has adopted an integrated national approach to environmental management. The FEPA's strategy within the framework of sustainable development is anchored in the following principles:

- sound policy formulation and implementation through both bottom-up and top-down coordination approaches;

- an effective environmental education and public-enlightenment drive;
- a strategy of local dialogue, consultation and cooperation and active international participation.

To fulfil its mandates to protect Nigeria's environment, the Agency has published *National Guidelines and Standards for Environmental Control in Nigeria*, which was issued in March 1991. This volume contains appropriate effluent-limitations regulations and pollution-abatement regulations for industries and facilities generating wastes. These control measures were launched in August 1991 to protect our water resources and other environmental resources. In addition, interim national water-quality standards and guidelines for various water uses were proposed in September 1991 during the international seminar on Water Quality Monitoring and Status in Nigeria.

In order to ensure compliance with its regulations, the Agency has established two departments—the Enforcement and Inspectorate Department and the Environmental Quality Department. Further, it has established the National Reference Laboratory/Zonal Office Complex in Lagos and proposed five Zonal Office/Laboratory complexes in different parts of the country. Each of these laboratories will be adequately equipped for water- and waste-water-quality monitoring. Within this framework, the National Data Bank for Environmental Monitoring and Management has undertaken the development of a water data bank component. Nigeria is also negotiating joining the GEMS/Water Network.

Moreover, FEPA is catalysing the implementation of all these activities through sectoral linkage and cooperation with other ministries and agencies that have mandates related to water resources and water quality so as to attain the national goal of sustainable development.

In conclusion, it is clear that Nigeria's resources have been abused by various national socio-economic activities, population pressure, rapid urbanization and a lack of environmental consideration in the country's decision-making processes. The health and wealth of the people of Nigeria is at risk and the nation cannot afford the huge price of failure as a spur to act. FEPA is calling for collective responsibility and cooperation among the government, industries, bilateral and multilateral agencies, non-governmental organizations and citizens in order to attain the goals of sustainable development.

DISCUSSION

Question: What is being done about sewage from latrines?
Answer: A new system of soakways is being built, especially in and around Lagos.

Question: Given the size of Nigeria, how can the country's water pollution be monitored?
Answer: There are five zonal authorities equipped with laboratories. Regional auth-

orities are being encouraged to comply with monitoring regulations. Not only have guidelines been distributed; they have been accompanied by a series of lectures.

Question: What is the penalty for violating regulations?
Answer: They vary with the offence, but they all involve imprisonment.

Chapter Forty-Three

Watershed management problems in Nepal: strategy for people's participation

RUDRA P. SAPKOTA
Department of Soil Conservation and Watershed Management, Ministry of Forests and Environment, Nepal

INTRODUCTION

Nepal is a small mountainous country covering an area of 147 487 sq. km which extends about 800 km east to west and about 120 km north to south. The total population of the country was 18.4 million in 1991 with an annual growth rate of more than 2 per cent. Over 90 per cent of the population is dependent on agriculture and related activities and contributes about 60 per cent of the nation's GDP. The intense pressure on its valuable but limited land resource-base has contributed to increasing absolute poverty, which is now more than 43 per cent.

Physiographically, the country has been divided into five regions generally based on ecological conditions from south to north. They are Terai (Plains), Siwaliks, Middle Mountains, High Mountains and High Himalayas.

Nepal's key environmental problems are associated with over- and under-use of the country's natural resources. The uneven distribution of human and natural resources, inadequate infrastructure facilities and the nation's lack of economic development are the basic causes of its environmental problems.

WATERSHED DEGRADATION

Watershed degradation is one of the major problems engendered by both natural and anthropogenic factors. Based on a recent classification of watershed conditions, one-third of the country's districts are considered to be marginal to very poor on average. Of a total of 75 districts, 25 are broadly categorized as in good condition, 25 fairly good, 13 marginal, five poor and seven in very poor condition. Natural factors causing watershed degradation are related to climate, geology and land form. The country gets most of its rainfall from the southeast monsoon, which ranges from over 3000 mm to 1400 mm annually, except in the trans-Himalayan

Table 1   Land forms and their characteristics in Nepal

Physiographic Regions	Percentage of Total Area	Altitude (m)	Geology
Terai	14	60– 330	Fine gravel
Siwaliks (stone)	13	200–1500	Coarse sand silt and stone
Middle Mountain (MM)	30	800–2400	Phyllite, quarzite, limestone
High Mountains (HM)	20	2200–4000	Gneiss, quartz, mica, schist
High Himalayas	23	above 4000	Gneiss, schist

Table 2   Land area, by dominant slope, in Nepal in degrees (approx.)

Slope	Area (sq. km)	Percentage
Steep to very steep	86 210	58.8
Moderate to steep	31 730	21.7
Dissected	1980	1.4
Gentle slope	6730	4.6
Very gentle slope	19 680	13.6
Total	146 330	100.0

zone of the northwestern region, where it is about 600 mm. Almost 80 per cent of the rainfall occurs within the short span of four months from June to September; the rest of the year is almost dry except in the west, where there is sparse winter rain.

The mountains of Nepal are very young and fragile and are highly susceptible to erosion. Another factor contributing to watershed degradation is rugged terrain and slope. The steeper slopes are most vulnerable in this regard.

However, Nepal does not have a serious wind erosion problem except in a small part of the trans-Himalayan zone in the northwest region.

The human activities responsible for watershed degradation are manifold and stem from socio-economic factors such as poverty, illiteracy, lack of economic development and employment opportunities other than in the traditional sectors.

LAND USE CHANGES AND DEFORESTATION

At present, 21 per cent of the land is under cultivation, 37 per cent is covered by forest and 17 per cent by grass and shrub. Out of the total cultivated land, about 50 per cent is in the Terai. Although in 1964 Nepal had about 6.4 million ha of forest, in 1985 that area had been reduced to 5.5 million ha; moreover, the highest densities have been reduced tremendously. These figures demonstrate that the depletion of forests is rapid. Currently, it is estimated that some 100 000 ha annu-

Table 3   Estimate of annual soil loss under various land uses

Land Use	Soil Loss (tons/ha/year)
Well-managed forest	5– 10
Well-managed terraces	5– 15
Poorly-managed sloping terraces	20–100
Degraded range lands	40–200

ally are being cleared by logging, encroachment and by fulfilling the needs for fuelwood and fodder.

The increasing pressure on limited cultivated land due to Nepal's rapidly growing population, coupled with unscientific farming, has caused such phenomena as over-use, shifting cultivation and farming on unsuitable slopes. The performance of the agricultural sector, with its increasing population/land ratio has resulted in a vicious circle of land degradation.

Another important factor that contributes to watershed degradation is livestock population. Nepal has one of the highest livestock densities per unit area of cultivated land. It has caused the overgrazing of pasture and other grazing lands and the removal of fodder from forests, resulting in the depletion of vegetation cover.

Some developmental activities aimed at providing basic infrastructure for economic development, such as hydroelectric dams and highways in the mountain and hill regions, have been detrimental to the resource base, many causing high soil erosion that adversely affects these structures themselves, agricultural lands and lakes downstream.

The nature of Nepal's energy sources explains in large measure the bleak picture of the country's forest depletion, as over 75 per cent of the nation's energy requirements come from fuelwood. Hence, deforestation cannot be controlled unless there is a reliable source of alternative energy, especially in the rural sector, which constitutes about 95 per cent of the population. Rural women are particularly conscious of the dwindling supply of fuelwood and fodder, as well as cheap alternatives, because collecting this material is a major part of the burdens they bear.

Other factors that contribute heavily to watershed degradation are unsound farming systems, uncontrolled grazing and the lack of integration among Nepal's development agencies.

SOME CONSEQUENCES OF WATERSHED MISMANAGEMENT

Soil loss

The loss of topsoil is one of the serious consequences of the watershed degradation. It is estimated that Nepal annually loses about 240 million m³ of soil which amounts to 1.5 cm for the entire country. Similarly, the sedimentation in the Terai river beds is about 15–30 cm/year, causing flooding.

Another major consequence of the watershed degradation is reduced crop pro-

Table 4 Annual loss of nutrients from rainfall

Nutrients loss (tons/ha/year)	Land use			
	Irrigated rice lands	Level terraces	Sloping terraces	Shifting cultivation
Organic matter	0	0.150	0.60	3.00
Nitrogen	0	0.008	0.03	0.15
Phosphorus	0	0.005	0.20	0.10
Potassium	0	0.010	0.40	0.20

duction, especially in the hills. Despite the high priority accorded to this sector, its performance declined and Nepal has recently become a net importer of food grains. For example, the average yield of maize and wheat (the second and third chief cereal crops after rice) was 1.8 mt/ha and 1.17 mt/ha in 1970, which decreased yet again to 1.6 mt/ha and 1.16 mt/ha, respectively in 1980.

Public participation

Lack of funds, inadequate institutional frameworks, ineffective policies and legal instruments, lack of coordination and public participation are the major management problems in Nepal. Noticeable improvements do not seem to take place except in integrated watershed management programmes, i.e. complete treatment of a watershed on an eco-unit basis except in a few cases. The Department of Soil Conservation and Watershed Management is the responsible body but so far has offices in only 27 districts out of 75.

STRATEGY FOR PUBLIC PARTICIPATION

Since watershed management requires the cooperation and popular participation especially of the farmers in the rural sector, some new strategies have been implemented by the government; in some cases these have had considerable success.

The planning process is both annual and periodic, the latter for five years. For the formulation of periodic plans, district authorities are given a budget ceiling for five years. Based on this ceiling and the management plan for programmes and priorities, they draft a periodic plan that will be discussed in the district assembly, which is composed of political representatives from all the villages within the district. In this assembly, the district officer asks for the contribution of the public in various programmes and, after review and whatever changes in programmes and priorities the members deem necessary, the assembly adopts the plan with the commitment of popular contributions (10 to 50 per cent of costs) and forwards it to the department.

Based on the periodic plan, the budget ceiling and the management plan, the district office sets its priorities, programmes and targets and prepares a draft annual plan. It again goes to the district assemblies for discussions. Priority is given to those villages which have volunteered the greatest participation and given sufficient assurance of meeting their commitment. Given these conditions, the individual project will be designed and implemented later on. Experience has shown that the share of public participation in terms of labour and materials ranges from as low as 10 per cent in the management of public lands to as high as 50 per cent in the treatment of private lands, one example being terrace improvement.

During the implementation and maintenance phases of the projects, a user group will be formed among those users who benefit from the specific projects. The district office provides technical advice and some funds to this group, which implements the project. In some cases, the district office designs and estimates the cost of the project and reduces the amount of its contribution, releasing these funds to the group according to the agreement both have reached. In cases where a contractor is involved, there are constant meetings between the user group and the district office to ascertain the villagers' contribution at the time their action is needed.

SUPPORTIVE MEASURES AND PUBLIC AWARENESS

To ensure maximum public participation, the district office launches some supportive measures that may not be directly linked to watershed management such as the promotion of cash-raising activities and skill development training. In addition, some programmes, such as those to improve the drinking water supply and school building maintenance, are conducted to satisfy the community's immediate concerns.

Creating awareness through various media, school programmes or prizes for the best farmers is considered a part of regular programmes to solicit popular participation.

Chapter Forty-Four

Energy and its environmental impact in Rwanda

MATHIAS NKULIYINGOMA
Agency for Environment and Development, Ministry of Planning, Rwanda

ASSESSMENT OF THE SITUATION

Problems caused by deforestation

Energy has always been and still is the driving force behind the development of countries and peoples. In all sectors, human activities need energy, be it in housing, industry, services, transport or any number of other endeavours. Energy consumption is currently one of the criteria applied in estimating the level of development of human communities.

Various studies indicate that fuelwood and plant wastes have a dominant place (96.6 per cent), in meeting Rwanda's energy requirements. Petroleum products meet only 2.9 per cent, while the remainder is met by local hydroelectric power, peat, imported electrical energy and imported gas.

The country's consumption of fuelwood is enormous, far above the level of production, with an annual deficit in the order of $2.3–2.8 \times 10^6$ m^3. The reason for this over-consumption of fuelwood is, above all, its use for cooking food on the traditional so-called "three-stone" (*amashyiga*) stove, which has a very low energy yield (approximately 5 per cent), as well as an efficient process of converting wood to charcoal. This deficit has led to over-exploitation of the wood resources provided by the natural forests, wooded savannas and tree plantations, with the following effects:

- a decrease in natural forests and a virtual disappearance of wooded savannas, except for Akagera National Park;
- the decrease in plant cover throughout the country with all that it entails in terms of climate irregularity and greater soil erosion.

Alternatives

An important alternative to wood as the main energy source is hydroelectric energy. Rwanda's topography and high levels of rainfall give it a significant hydraulic potential that could be converted into electricity. A serious obstacle encountered in supplying rural areas with electricity on a large scale is the scattered nature of the rural population. Other alternative energy sources in use on a small scale in Rwanda are biogas, methane gas, peat, papyrus and solar energy. Nevertheless, the amount of hydrocarbons imported will remain high because 77 per cent of such imports are used in transport, a sector of the economy that is expected to expand greatly.

STRATEGY

It is clear that the very high consumption of wood, as a result of high population density, low-performance cooking techniques and charcoal production, is the principal cause of wood shortages and diminishing woodlands. The general objective of the environmental policy relating to energy is the achievement of a sustainable level in the use of wood resources, consumption of which must no longer exceed the rate of renewal, by encouraging the use of other non-polluting forms of energy (hydroelectricity, gas, solar energy), and by taking measures to increase energy yields.

The specific objectives with regard to energy are:

1. To develop energy sources as alternatives to fuelwood and plant waste, renewable and less polluting forms of energy sources;

2. To increase energy yields by improving cooking and charcoal production methods;

3. To improve the access of rural people to alternative forms of energy (electricity, gas);

4. To promote the use of electricity instead of fuelwood;

5. To pursue research on the development of new and renewable energy sources.

PLAN OF ACTION

Ongoing activities to give effect to environmental policy relating to energy to include:

1. Energy infrastructure and equipment projects intended to improve access to sources of energy other than fuelwood and, as a direct consequence, reduce the consumption of this resource;

2. Research, development and extension projects aimed at reducing wood consumption popularizing the use of alternative forms of energy;

3. Projects to renew and preserve wood resources.

Priority activities include:

- the development and dissemination of higher-performance techniques and equipment for charcoal production, as the use of charcoal is more widespread in urban areas than in the countryside;
- bringing people together in larger settlements to facilitate their access to energy distribution networks (electricity, gas);
- creation of production units for improved stoves.

The large number of projects and the diversity of operators in the energy sector is proof of the importance accorded to it in Rwanda.[*]

*Editors Note

The environmental effects of civil war

Most of the strategy and projects described in this chapter have been overtaken by the events of 1994. In particular, there has been dramatic devastation of forests for fuelwood around Kigali and elsewhere.

Chapter Forty-Five

The Mount Apo National Park Geothermal Project: reconciling imperatives in national development

MARIO LUIS J. JACINTO
Department of Environment and Natural Resources, Philippines

Past energy projects, the world over, have generated socio-economic and environmental problems that seriously threaten the implementation of proposed energy facilities, despite precautionary measures to avoid repetition of such developments. This is the dilemma facing the Mount Apo National Park and Philippine National Oil Corporation (PNOC) Geothermal Project. The experience of communities with energy projects like the Tiwi Geothermal Power Plant has given rise to a generally negative attitude on the part of the people affected, stemming mainly from the perception that, although they are made to carry the burden of environmental and social costs, they benefit little or not at all from these undertakings. This is particularly true in the case of the Tiwi project, where expectations of economic benefits trickling down to the local population were never fulfilled. Instead, the affected populace was forced to bear the burden of the environmental and social costs generated by the plant such as water and air pollution and slow payment for their expropriated properties. The impression of inequitable distribution of the fruits of development has therefore been ineradicably etched into the minds of local people. Unfortunately, there is no existing policy or regulatory framework designed to mitigate the brunt of these socio-economic costs. Hence, although every agency is aware of the problem, they could not immediately undertake ameliorative measures.

THE PROJECT

The Mount Apo Geothermal Project is being developed to respond to the increasing demand for power supply in Mindanao and also to ease further the Philippines' dependence on oil-based generated power and consequently, to decrease the country's vulnerability to the imported oil supply–demand problem. Mount Apo is also one of the seven sites identified by the Department of Environment and Natural Resources (DENR) as pilot areas for the Integrated Protected Areas System (IPAS).

The proposed development of the Mount Apo Geothermal Project resulted from very encouraging results of exploration conducted from 1986 to 1988. Resource assessment calculations estimate available geothermal power as sustainable for 25 years at a minimum of 292 MW and a maximum of 548 MW.

The project is situated at the northeastern flank of Mount Apo, which is considered the Philippines' highest peak at 2954 m in Kidapawan, Province of North Cotabato. All development activities will be confined to a rectangular area of 701 ha; only 0.2 per cent of the Mount Apo reservation's 52 262 hectares will be opened up.

The field development and operational phases of the project will cover the following major activities:

- construction of an additional 6.7 km of road network;
- drilling and testing of 23 production wells and seven reinjection wells located within seven production pads and two reinjection pads, respectively;
- construction of and cooperation in a single power plan of 6×20 MW capacity.

THE ENVIRONMENTAL IMPACT ASSESSMENT REVIEW PROCESS

The Environmental Impact Assessment (EIA) Report was submitted in February 1989. However, due to comments on the need to improve on the EIA (as pointed out by interested parties in dialogue, including a senate hearing held in Davao City on 20 May 1989), a decision was taken to conduct more studies and improve on the EIA report. The cabinet also helped consolidate issues and answers on the project by creating a working committee on 16 May 1990.

To make sure that the final EIA Report would be complete and focused on critical issues, the Environmental Management Bureau of the Department of Environment and Natural Resources drafted guidelines with National Energy Office (NEO) counterparts on 20 June and 18 July 1990. After about five months of additional studies and consultations, including two formal tribal consultations, the PNOC submitted the EIA report to the DENR Environmental Management Bureau.

Unlike other projects, a special EIA Review Committee was recruited from academics, research and other institutions and local and national officials. Special care was taken that the most credible and unbiased experts on the issues raised formed the Review Committee.

The public hearing on the project scheduled for 12 February 1991 was converted into a consultative meeting because of a technicality concerning announcements. A second public hearing was held in Kidapawan on 10 April 1991. After this public hearing, meetings by the Review Committee with PNOC officials (22 April 1991) and with opposed groups such as Task Force Sandawa (8 May 1991) continued. The EIA Review Committee submitted its recommendation on the project to the DENR on 15 May 1991. The whole process, because of the intensive studies required and the many consultations, hearings and meetings to assure the correctness of decisions, took more than 14 months after the submission of the first EIA report.

THE ROLES OF VARIOUS GROUPS

The EIA consultants

The first EIA report submitted was largely written by PNOC's "in-house" consultants. However, because of the strong demand to improve on this EIA, the PNOC was asked to organize an outside group to be composed of the best experts on the various concerns related to the project. Most of the experts are based at the University of the Philippines and are well-known to be pro-environment. The wildlife expert is a stalwart of Haribon, University of the Philippines, Los Baños. The anthropologist had done research on the Lumads (a tribal group) for many years, even before the issue arose and was a strong critic of the project as originally planned.

The cabinet

The cabinet, through a working committee especially organized for this project, facilitated exchange of information between relevant groups and agencies and helped consolidate opinions from different perspectives. To the president's credit, she did not use the authority given her under the EIA Law to exempt the project from the EIA requirements in the name of national development. She did not "short-cut" the process but instead gave full play to the democratic process that forms the basis of the EIA system as DENR wants it implemented.

The DENR

In this case, the Environmental Management Bureau of DENR undertook the most intensive EIA review of any project in the history of EIA implementation in the country. The EIA study was pursued and carefully formulated by DENR–EMB in consultation with concerned NGOs, academics, tribal representatives and the project proponent. The decision to deny or approve the application for an environmental compliance certificate became a collegial decision of concerned environmental experts, who formed the Review Committee, as well as citizens who participated in public hearings and consultations.

THE EIA REVIEW COMMITTEE RECOMMENDATIONS

The Committee recommended the following conditions:

1. Only a single power-generating plant with 120 MW capacity was to be developed.

2. Only a total area of 111.57 ha would be opened up as permanent development areas for wells, roads, pipeline corridors, power plant and ponds and other facilities to be used during the construction period.

3. The PNOC was to implement the watershed management plan as submitted by the Environmental Management Bureau.

4. All necessary permits from other government agencies would be secured.

5. There was to be no discharge of geothermal effluents and solid wastes during any phase of development and power generation.

6. The ambient air quality emissions, including noise, would be kept within the prescribed standards.

7. Risk assessment studies and a corresponding contingency plan were to be submitted within 60 days of the issuance of the environmental clearance certificate.

8. Appropriate tribal propitiatory rites, such as *pamaas*, were to be performed as long as they were legal and non-violent.

9. An environmental and forest protection guarantee fund was to be established.

Other conditions include giving qualified members of affected cultural communities preference and priority in employment for the project. In cases where positions are not readily available to the affected tribal members, the same positions will be made available to the immediate next of kin of these families who are qualified for the job. All displaced families are to be relocated to an area equivalent to that which they vacated. Health and education services shall be provided by the PNOC, and the existing tribal councils covering the affected families shall be reorganized to constitute the Tribal Consultative Council.

CONCLUSION

The project has thus been approved, adopting the recommendations of the Review Committee including the following major condition and restriction to safeguard ecological and social imperatives:

1. Royalties will be paid to the affected upland dwellers who regard the site as their domain and sacred ground;

2. Only a single power-generating plant of 120 MW shall be developed.

The Mount Apo Geothermal Project provides the opportunity for the government to reconcile three equally important thrusts: environmental protection, energy sufficiency and the recognition of the rights of indigenous communities in their ancestral lands. It has also brought about policy reforms in terms of equitable access to resources and benefits from government projects. The requirements gave credence to hopes to balance the concerns of development with environmental objectives, which will be reinforced through the involvement of multi-sectoral organizations concerned with the protection of the Mount Apo National Park.

DISCUSSION

Question: How are the tribal groups affected by the project being protected?
Answer: A multi-sectoral body called the Mount Apo Protection Committee was formed, in which their leaders were included, along with representatives of the local governments concerned and the Ministry of the Interior. The 53 families that had to be resettled were compensated and given first preference for employment.

Question: How will the biological diversity of the area be preserved?
Answer: Some 700 ha of the area have been set aside as a protected area.

Chapter Forty-Six

Information networks for energy planning: the role of the UNESCO Regional Office for Science and Technology in Africa (ROSTA)

A. ABDINASER
Former Acting Director, UNESCO/ROSTA

INTRODUCTION

In many developing countries, a double energy crisis has emerged as a result of the sharp increase in the real price of oil during the past few years and a growing shortage of traditional fuels such as wood. Consequently, attention is increasingly being paid to possibilities for enhancing both the importance and use of new and renewable energy sources.

ENERGY POLICY FORMULATION AND PLANNING

The energy component of the Lagos Plan of Action, including the development and exploitation of both new and renewable sources of energy and non-renewable sources of energy, will have to be determined at the national and multinational levels. Common to all aspects of energy problems in Africa are fundamental weaknesses and deficiencies that require concrete and urgent rectification at the national, sub-regional and regional levels. These deficiencies include:

1. the absence of national energy policies and development programmes integrated into national development plans;

2. the insufficient capabilities, human and institutional, for drawing up and using inventories of all energy resources, particularly of non-renewable sources;

3. the lack of capability for planning and the composite exploitation of all available energy resources, taking into account projected development by different sectors, as well as technology changes for the exploitation of different forms of energy;

4. the lack of adequate personnel for the evaluation, extraction and processing of all forms of energy for the design (including standardization), manufacture and marketing of equipment components and spare parts for research and development;

5. the tendency to use inappropriate technologies from industrialized countries;

6. inconsistencies in policies, planning and programme and project designs and execution, and a lack of information at the national and multinational levels on achievements in the development and utilization of non-renewable energy resources, the possibilities of obtaining equipment using such sources, its limitations, etc.;

7. the need to orient energy development policies—and therefore the mobilization and redeployment of financial resources in favour of small-scale development projects, especially to meet the needs of rapid development in the rural sector;

8. the inadequacy of cooperation between technical institutions concerned with resources inventory and planning, and joint feasibility studies and multinational enterprises responsible for production, standardization and marketing, research and development, human resources, development, market surveys, etc.

In the African countries, energy policies and planning must take into account the socio-economic duality created by the co-existence of modern and traditional sectors. This duality is characterized by different patterns of energy consumption and supply: in the modern sector, life style and technologies resulting in energy use largely based on petroleum; in the traditional sector, a subsistence economy based on low-energy intensity, dispersed demand and supply but often a threateningly high consumption of limited fuelwood resources which leads to the accelerated degradation of ecosystems.

Obtaining information on the recent research, application, case studies, manufacturers, expert consultants or patents concerning alternative energy sources is not easy. This information has been collected for only a few developed countries, sometimes in only one field of alternative energy.

There is very little consolidated energy information either on or available to developing countries as a whole. Energy information is provided, in most cases and particularly in Africa, by a wide variety of institutions and individuals in many fields, including the physical and environmental sciences, the life sciences, engineering and economics. This diversity creates problems for locating, evaluating and using such information. At present, some developing countries lack even the framework and resources to collect information that could be useful to them. Ways of channelling information to groups of users are often inadequate. The absence of standardization also prevents comparisons of standards.

THE CONTRIBUTION OF UNESCO

UNESCO has taken up the challenge of helping to manage the world's energy information, especially as regards new and renewable sources of energy for developing countries. The purpose of its Energy Information Programme is to put the right information in the right hands at the right time. Launched in January 1981, the programme is conducted in association with UNESCO's General Information Programme. This deals with scientific and technological information, libraries and archives and the promotion of information systems and services at the national, regional and international levels and includes the Intergovernmental Programme for Cooperation in the Field of Scientific and Technological Information (UNISIST).

UNESCO is also supporting a Cooperative International Network for Training and Research in Energy Planning (CINTRE), which aims at the establishment of a network of stable interdisciplinary and interactive academic centres in developing countries. These would offer training activities at different levels and carry out research on the technical, economic, social environmental and related problems of the energy systems of the countries or region. While only one African Centre (ENDA) is currently participating in this Network, it is hoped that other African centres will participate in the future.

One of the major programme activities of UNESCO concerning information on energy is the launching of regional projects for enhancing existing information services and encouraging the sharing of information resources in an effort to link organizations in an international network. Pursuant to this overall objective, sub-regional and regional pilot projects are being undertaken in five regions of the world. In Africa, two sub-regional pilot projects on non-renewable energy resources information-sharing networks have been established by UNESCO: one for eastern and southern Africa and the other for west Africa.

An African Regional Symposium on Solar Energy was organized by UNESCO in November 1983 in Nairobi, Kenya. This Symposium was a prelude to the formation of the Solar Energy Society of Africa (SESA).

DISCUSSION

Question: Why aren't greater resources devoted to renewable energy sources?
Answer: This field calls for a long-term approach and perspective. Many universities, research institutes and comparable bodies tend to allocate their funds to areas in which rapid results can be obtained.

Question: What are the main alternative energy sources whose development you would encourage in Africa?
Answer: Solar energy, biomass and hydroelectric power. The River Zaire, for example, has enormous hydroelectric power potential. The development of solar energy for the continent is particularly appropriate. There are a number of initiatives now under way to improve solar captors to produce hot water and steam, as well as major efforts to reduce the cost of photovoltaic cells.

Comment: In Zimbabwe, the cost of energy produced by photoelectric cells has plunged from US$ $200 to $0.20 per unit.

Question: Since so many women in developing countries are involved in income-generating activities based on energy inputs, to what extent do women participate in UNESCO training activities in the field of energy?

Answer: UNESCO strongly encourages women to participate in training activities concerning technology and energy. In fact, UNESCO's three current priorities are: first, women; second, Africa; and third, the least-developed countries.

Chapter Forty-Seven

The UNEP Industry and Environment Programme Activity Centre (IE/PAC)

KIRSTEN OLDENBERG
IE/PAC, UNEP

UNEP's Industry and Environment Programme Activity Centre, based in Paris, was created in 1975 to bring together industry and governments to work towards environmentally sound development. The Centre acts as a catalyst; most of its activities are performed in conjunction with the United Nations Industrial Development Organization (UNIDO), the International Labour Organization (ILO), the United Nations Development Programme (UNDP), the Organization for Economic Cooperation and Development (OECD) and the International Chamber of Commerce (ICC), as well as national chambers of commerce, corporations and lending organizations. Basically, the Centre's work falls into three basic categories:

- *Training and technical cooperation,* one recent example of which was the 1991 regional workshop for Latin American and Caribbean countries on landfill of hazardous industrial wastes;
- *Information transfer,* facilitated by the quarterly *Industry and Environment Review* and a query-response service that in 1991 answered more than 3000 requests by 74 countries on subjects ranging from agro-industries to tourism and transport; and
- *Publications,* notably a series of technical reports such as the recent *Audit and Reduction Manual for Industrial Emissions and Wastes* (published jointly with UNIDO) and other publications such as the cooperative venture with the International Petroleum Industry Environmental Conservation Association, *Climate Change and Energy Efficiency in Industry.*

Half of the Centre's staff, working under the Directorship of Jacqueline Aloisi de Larderel, this contributor included, are seconded from governments and industry, a fact that underscores concern in both the public and private sectors about environmentally viable industrial development.

The Centre's four current major programmes are the following:

- *Awareness and Preparedness for Emergencies at the Local Level* (APELL), which merits a chapter in itself;
- *Ozonaction*, undertaken in conjunction with the Interim Multilateral Ozone Fund of the Montreal Protocol on Substances that Deplete the Ozone Layer, which, through information exchange, regional workshops and country programmes, helps developing countries replace CFCs and halons in industries such as refrigerants, solvents and sterilants;
- *Tourism and Transport*, the former carried out in close cooperation with UNESCO and the World Tourism Organization (WTO) on the development of the World Heritage Sites, the latter with the World Bank on such issues as environmental impact assessment and environmental management of roads and automotive air pollution; and
- the Cleaner Production Programme described below.

The Cleaner Production Programme essentially involves a rethinking of the entire production process in virtually every industry. We now tend to think in terms of "end-of-pipe" processes—essentially *reactive* strategies for disposing of production wastes, including their detoxification or recycling. Cleaner production involves a shift to *preventive* action. We now know that our capacity to manipulate our environment has far outstripped our ability to understand the effects of that manipulation—the destruction of the ozone layer by CFCs and halons being a perfect example of our limited comprehension of the ultimate impacts of our activities. Neither CFCs nor halons are toxic in themselves. We simply failed to grasp the impacts of their uses.

It is useful to think of cleaner production as a continuum of questions, all of which must be asked again and again each time the slightest change in technology modifies the production process in any field in any way:

Can the production of a by-product be prevented?
Can it be re-used?
Can its re-use take place off-site or on-site?
Can it be treated (detoxified) or destroyed?
Must it ultimately be buried?

One can—and should—apply this hierarchy of questions to the household as well as the production plant. It concerns such daily chores as washing dishes or clothes, to say nothing of cooking. In short, as it has been officially described: "Cleaner production means the continuous application of an integrated preventive environmental strategy to processes and products so as to reduce the risks to humans and the environment."

For *processes*, cleaner production involves conserving raw materials and energy, eliminating the use of toxic substances and reducing the quantity, as well as the toxicity, of all emissions and wastes before they leave any given process. For *products*, it means reducing their impacts during their entire life cycle, from raw material extraction to disposal.

Cleaner production is achieved by applying knowhow, improving technology and above all, *changing attitudes and habits*. It needn't be expensive. One doesn't necessarily need new plants. It may simply involve changing valves in order to prevent leaks or eliminating processes that mix water with wastes, if only to reduce the volume of water. And the cost-benefits of cleaner production are immense, notably in reduced energy expenses and improved efficiency. Indeed, end-of-pipe-line strategies are far more expensive: in the United States alone, US $5 billion is spent annually to clean up the results of dirty production.

One doesn't even need a regulatory framework and strict enforcement procedures to achieve cleaner production. This is simply the Western experience—and a very expensive one it is. Instead, the governments of developing countries can use a variety of economic incentives, such as tax reductions, to encourage cleaner production within their borders. What is needed—by *all* countries—is research and development aimed at new technologies, as well as a great deal of information.

On this latter point, the Industry and Environment Programme Activity Centre stands ready to help. Its computer-based information exchange system, the Cleaner Production Information Clearing-House, known by its acronym ICPIC, can provide not only single answers to single questions, but 600 case studies and 400 publications. It also draws increasingly on information furnished by several cleaner production working groups, ranging from metal finishing to textiles, and publishes a semi-annual newsletter that supplements the *Industry and Environment Review*. At the request of governments, the Centre can and has held seminars and can help draft Cleaner Production Action Plans for a country or region.

It is often said that three factors contribute to the cost of a house: location, location and location. There are three comparable determinants of cleaner production: education, education and education—at all levels from primary school onwards for all members of society.

DISCUSSION

Question: Can the UNEP Industry and Environment Programme Activity Centre train inspectors on change in industrial technology?
Answer: Yes. But formal training by the Centre may not be necessary. The Centre has produced, in collaboration with the Environmental Education and Training Unit and other units of UNEP, a number of training manuals.

Question: What was the focus of the recent UNIDO Conference on Ecologically Sustainable Development (held in 1991)?
Answer: Cleaner industrial production. One conclusion was that cleaner production protects the environment far more inexpensively than end-of-pipeline approaches.
Comment: Developed as well as developing countries need to produce environmentally friendly goods.

Answer: Absolutely. For the moment, the Industry and Environment Centre is focusing on developing countries because their needs are greater—notably those that are newly and rapidly industrializing, especially in southeast Asia, where the process of industrialization is generating numerous and immense environmental problems.

Question: Can the Centre help countries with the environmental problems created by the reactivation of old mines and the techniques of surface mining?
Answer: We hope to. There are certainly ways of decreasing the damage caused by surface mining. But we are only beginning to approach mining projects.

Chapter Forty-Eight

Awareness and preparedness for emergencies at the local level (APELL)

KIRSTEN OLDENBERG
Programme Officer, Industry and Environment Programme Activity Centre,
UNEP

The propane explosion in Mexico in 1984, methylisocyanate rampant at Bhopal that same year, the fire and discharge of contaminated waters into the Rhine from a Basel warehouse in 1986, Chernobyl ... emergencies such as those led UNEP to outline a cooperative approach to technological accidents in 1988 with the support of the international chemical industry, which is called APELL, with all its reverberations in the French language of a call. This programme for Awareness and Preparedness for Emergencies at the local level has twin basic aims: *to prevent* industrial accidents and *to minimize their impacts*. In some ways, it is an extension of cleaner production.

Similarly, the APELL process involves two fundamental activities.

- to *create* and/or increase community awareness of the risks involved in the manufacture, handling and use of hazardous materials—and of the steps taken by authorities and industry to protect the community from them;
- using this information to *develop*, in cooperation with the local communities, emergency response plans involving the entire community.

We might as well call these two elements "Community Awareness" and "Emergency Response". The key word is *community*, for the APELL process, properly worked out, creates a community of interest among the local (or provincial) authorities, industrial plant managers and their work forces, and finally, the citizens at large, including interest groups concerned with health, the environment, the media, education, business and last, but hardly least, religious establishments.

The key organizational step towards making the APELL process work is the formation of a Coordinating Group representing the various constituencies that must have a voice in the establishment of an emergency response plan. These are basically representatives of the local authorities, local community leaders and industry—any one of which may launch the APELL process, making sure that any and all groups with an interest in the planning process are included. The plant managers of the area's industrial facilities also must assure the local authorities and the community leaders that they are acting on behalf of the highest officials of their enterprises.

Initiating the APELL process best suited to the particular locality involves essentialy ten steps:

1. identifying the emergency response participants and establishing their roles, resources and concerns;

2. evaluating the risks and hazards for the community;

3. reviewing the emergency plan of each participant or group of participants for its adequacy in relation to a coordinated response;

4. identifying the required response tasks not covered by existing plans;

5. matching these tasks to the resources available from the participants identified;

6. making the changes necessary to improve existing plans; integrating them into an overall community plan; and securing the agreement of all the participants;

7. committing the integrated plan to paper as a formal written document and securing the assent of all those in the local (or provincial) government services who will be involved in the plan's implementation—the police, the fire service, the local health facilities;

8. educating all the participants about the integrated plan and ensuring that all emergency respondents are trained for its application;

9. establishing procedures for periodic review, testing and updating of the plan. As in the case of the cleaner production continuum of questions, the APELL plan must be rethought and reshaped each time a significant new element enters the process or, conversely, disappears from it;

10. educating the community at large about the plan and asking for their criticisms and contributions.

APELL is simply a guideline, for adaptation by communities. An APELL handbook is available in twelve languages, an APELL newsletter appears as a regular supplement to the *Industry and Environment Review*, an *International Directory of Emergency Response Centres* was recently co-published by UNEP and the OECD, and a number of APELL seminars and workshops have been held throughout the world. These are all aids to the development of individual APELL programmes—and more will follow in the months and years to come. But the basic responsibility of creating a concrete APELL programme lies with the community itself.

The bridges of responsibility among local industry, the APELL coordinating group and the local (or provincial) government is illustrated below:

INDUSTRY RESPONSIBILITIES	COORDINATING GROUP BRIDGING ACTIONS	LOCAL GOVERNMENT RESPONSIBILITIES
1. Assure safe work practices.	1. Open lines of communications.	1. Provide a safe community.
2. Assure personal safety of employees and visitors.	2. Information sharing.	2. Assure the safety and well-being of all residents and transients within community.
3. Establish safety programmes.	3. Coordinate emergency plans and procedures.	
4. Protect lives and property on-site.	4. Interact with other emergency response agencies.	3. Establish public safety programmes.
5. Coordinate all plant personnel during an emergency.	5. Joint education and common problem solving group.	4. Protect lives, as well as private and public property.
6. Develop plans and procedures to respond to emergencies.	6. Mutual aid assistance.	5. Coordinate community emergency response forces during an emergency.
7. Provide security, safety equipment, training and information on chemical hazards.		6. Develop plans and procedures to respond to emergencies.
		7. Conduct training, drills and exercises with other response agencies within the community area, or state.
		8. Maintain communication channels with national governments.

Chapter Forty-Nine

Hazardous wastes: their transboundary movement and disposal

IWONA RUMMEL-BULSKA
Coordinator, Interim Secretariat for the Basel Convention, UNEP

INTRODUCTION

International trade in hazardous wastes is a pernicious new form of commerce that afflicts the least-developed countries in particular. Air and water-borne pollutants move across borders largely because of inadequate technologies to halt them. Potential poisons, ranging from solvents to pesticides, are freely sold and bought for their potential benefit. But profiteering from the dumping of wastes that are not properly recycled or cannot be—by-products so lethal that their disposal is forbidden in their countries or regions of origin—became in the early 1980s an obvious arena of UNEP action.

The international traffic in hazardous wastes had then become big business, consisting of more than 20 000 cross-frontier movements annually from Western to Eastern Europe alone. Such shipments from North to South amounted to some 10 per cent of *total* transboundary movements, the more serious because these cargoes of poison were and continue to be exported to developing countries that have no environmentally sound waste disposal facilities. In 1988, the media characterized such dumping as "garbage imperialism" and "toxic terrorism". Strangely enough, many of these shipments take place in accordance with national legislation or relevant international legal instruments. Such are the differences in national regulations, including the very definitions of what constitutes a hazardous waste, that this traffic was and remains extremely difficult to monitor and creates enormous possibilities by which the growing number of waste brokers can legally continue to operate.

THE BASEL CONVENTION ON THE CONTROL OF TRANSBOUNDARY MOVEMENTS OF HAZARDOUS WASTES AND THEIR DISPOSAL

The negotiation process that was to culminate in the elaboration of the 1989 Basel Convention on the Control of Transboundary Movements of Hazardous Wastes and Their Disposal began in 1981 in Montevideo, where the *Ad Hoc* Meeting of Senior Government Officials Expert in Environmental Law, created by UNEP's Governing Council, outlined issues of particular importance. One area so identified was the transport, handling and disposal of toxic and dangerous wastes. The following year, the Council established an *Ad hoc* Working Group to study this problem further. By 1985, after three meetings, the Group produced the Cairo Guidelines and Principles for the Environmentally Sound Management of Hazardous Wastes. These Guidelines, taken together with work done earlier by the Organization for Economic Cooperation and Development and the European Economic Community, constituted the foundation of the Basel Convention.

The basic aim of the Convention is to encourage countries to cut back on the quantity and toxicity of the wastes they generate, to manage them in an environmentally sound way and to dispose of them as safely and as close to their source of generation as possible. The treaty defines hazardous and "other" wastes (basically household wastes and incinerator ash) in a series of technical annexes, which include wastes subject to certain recycling operations. Consequently, the only wastes excluded from the Convention's scope are radioactive wastes covered by other international instruments. The Convention stipulates that countries should export wastes only if they cannot dispose of them safely within their own borders. It also ensures that any *authorized* traffic is carefully controlled.

In addition, the treaty stresses that each country has the sovereign right to ban imports of hazardous wastes entirely and that a party may not exchange wastes subject to the Convention with a non-party. However, parties may enter into bilateral, multilateral or regional agreements that must be no less stringent than the provisions of the Basel Convention. One such agreement is the Bamako Convention, to which African states are signatories.

Where transboundary movements of hazardous wastes are permissible, the Basel Convention provides for an elaborate control system based on the principle of Prior Informed Consent (PIC). Exports must obtain the prior written consent of the importing country before shipment. The countries through which the wastes will pass must also provide prior written consent and all handlers must possess a valid, officially issued document. Parties must also prosecute as criminals those responsible for illegal traffic. Additionally, the Convention requires industrialized countries to assist developing countries in improving their management of their own wastes.

CONCRETIZING THE CONVENTION FURTHER

The Basel Convention is hardly an end in itself. It is a framework for reducing the hazardous waste problem, a framework that requires many additional elements already undertaken by the Convention's Secretariat to make the treaty a far stronger instrument. Among these elements are the following:

● the development of a protocol on liability and compensation for damage arising from transboundary movement and disposal of wastes;

- the preparation of technical guidelines for the environmentally sound management of wastes subject to the Convention. These would provide information on the control expected by parties concerning the management of hazardous wastes produced within their territories and would include such subjects as:

1. preventing and managing waste generation;

2. identifying and characterizing hazardous wastes;

3. recycling, treatment and disposal facilities; and

4. the capacity and capabilities of the authorities concerned.

Such guidelines would also furnish advice to the authorities designated by the parties in making decisions as to whether to accept or reject proposed transboundary shipments going into or out of their territories.

Moreover, the Convention's Secretariat is studying the provision of temporary financial assistance in the case of emergencies to minimize damage from accidents arising from transboundary movements of wastes and their disposal. It has laid the groundwork for increasing public awareness of hazardous waste issues and promoting the dissemination of information on the subject, as well as developing education and training programmes for decision-makers in industry and government. Lastly, it has begun shaping strategies for enhancing cooperation among parties for the exchange of information on illegal transboundary movements.

FURTHER WORK IN THE LIGHT OF AGENDA 21 OF THE UNITED NATIONS CONFERENCE ON ENVIRONMENT AND DEVELOPMENT

The International Secretariat of the Basel Convention has before it a full programme deriving from the recommendations of UNCED's Agenda 21 concerning additional work on hazardous wastes that fall within the scope of the treaty. Already the Secretariat has begun providing assistance to parties in assessing the health and environmental risks that result from exposure to hazardous wastes and in identifying priorities for controlling the various categories of wastes. Also in conjunction with governments and the international organs concerned, it is planning to promote centres of excellence for training in hazardous waste management.

Further, it is contemplating a variety of activities to be undertaken with other units of UNEP, as well as other United Nations' bodies or interagency groups. The following represent but a few:

- developing guidelines for estimating the costs and benefits of various approaches to the adoption of cleaner production procedures, waste minimization and environmentally sound management of hazardous wastes, including the rehabilitation of contaminated sites;
- assisting developing countries in establishing hazardous waste treatment and disposal sites, as well as promoting the clean-up of contaminated areas;

- assessing the feasibility of establishing and operating national, sub-regional and regional waste treatment centres; and
- developing and expanding international networking among concerned professionals to maintain a continuous flow of information among parties.

This is an ambitious and arduous programme, the more so when taken in conjunction with other proposals that the International Secretariat is currently mooting. But it provides a few means of mitigating some of the scourges left to generations to come by a throwaway century.

Chapter Fifty

Toxic chemicals: the UNEP international register

JAN HUISMANS
Director, International Register for Potentially Toxic Chemicals (IRPTC),
Programme Activity Centre, UNEP (now Assistant Executive Director,
Earthwatch, UNEP)

In a world in which 1000–2000 new chemicals enter the global market each year, their possible dangers unknown, a world-wide information system on their attributes and potential effects is blatantly imperative. One has only to think of Minimata disease in Japan, the 1984 gas leakage at Bhopal and the million-odd people the world over annually poisoned by pesticides. Established in 1976 as a branch of Earthwatch, the International Register of Potentially Toxic Chemicals (IRPTC) attempts to bridge the gap between the prospective buyers and users of these substances and knowledge of their harmful characteristics, information often buried in an immense spectrum of specialized journals and reports.

The Register's data bank contains detailed profiles on more than 800 chemicals of international concern—their health and environmental hazard and risk assessment evaluation from their production through their use and disposal. Its Legal File compiles critical information on national and international regulations governing some 6000 chemicals. If IRPTC cannot directly supply the answer to a query, it calls upon the resources of a vast network of international organizations, scientific institutions and industrial contacts. In addition, more than 100 countries have appointed national correspondents who relay news on their latest research and laws to the Register. Much of this material is regularly disseminated internationally through an active publication programme, including a quarterly bulletin.

The four major functions of the Register are the following:

- making data on chemicals readily accessible to those who need it through its query/response service;
- locating and publicizing major gaps in this information, encouraging research to fill them;
- identifying and publicizing the potential hazards of using particular chemicals; and

- compiling information on national, regional and global policies for regulating the production and use of hazardous chemicals.

The Register serves both developed and developing nations. For the industrialized countries, it provides information on current research and international regulations and, in addition, draws attention to future chemical risks. Simultaneously, IRPTC offers developing countries access to research they could not afford to conduct themselves and thereby furnishes them with means of avoiding some of the mistakes made by their rich counterparts. IRPTC also helps developing countries to set up their own national registers.

Further, IRPTC has acted to protect these countries by ensuring that chemicals banned or restricted in developed nations are not dumped on their poorer neighbours. In 1984, following the adoption of a provisional notification scheme, UNEP's Governing Council approved the Amended London Guidelines for the Exchange of Information on Chemicals in International Trade, which includes a procedure for Prior Informed Consent (PIC) in the importation of a given substance. PIC has been applied to chemicals banned or severely restricted by some 20 countries. IRPTC informs other countries of these bans and offers guidance or training on possible action so that a government may decide whether or not to forbid the import of a particular substance. Subsequently, the Register relays this information back to the exporting countries.

THE INTERNATIONAL PROGRAMME ON CHEMICAL SAFETY (IPCS)

Launched in 1980 as a joint venture of ILO, WHO and UNEP to assess the risks posed by particular chemicals to human health and to the environment, the International Programme on Chemical Safety (IPCS) brings scientists together to harmonize research methods for comparable, reliable results; trains experts from developing countries in chemical control and assists governments in developing the machinery to cope with chemical emergencies, an activity to which the APPELL process, developed by UNEP's Industry and Environment Programme Activity Centre, contributes. IRPTC supplies the raw data for decisions on chemical safety, while IPCS, headquarters at WHO, processes this information, publishing evaluations of health and environmental risks, as well as advice on control measures and exposure levels, feeding such material back to IRPTC to enhance its data profiles.

IPCS issues its materials in four forms:

- the detailed *Environmental Health Criteria* documents for specialists;
- short, non-technical *Health and Safety Guides* for managers and decision- makers, which also provide advice on such practical matters as safe storage, handling and disposal of the particular chemical, as well as accident prevention and health protection measures;

- *International Chemical Safety Cards* for worker reference on site; and
- *Poison Information Monographs* for medical personnel.

THE PANEL OF EXPERTS ON ENVIRONMENTAL MANAGEMENT FOR VECTOR CONTROL (PEEM)

IRPTC also works closely with FAO in implementing that agency's 1986 International Code of Conduct on the Distribution and Use of Pesticides, notably through the UNEP/FAO/WHO Panel of Experts on Environmental Management for Vector Control (PEEM), founded in 1980. The Panel's work to promote alternatives to pesticides in controlling vector-borne diseases has become all the more critical as one after another agricultural pest has developed resistance to one or more pesticides. To date, work has focused largely on those aspects of water resources development associated with disease vectors.

LOOKING TO THE FUTURE

As the chemical industry grows almost exponentially, IRPTC must continue to expand and accelerate its current work, particularly, as recommended by the United Nations Conference on Environment and Development, through the strengthening of IPCS. The Register's role in implementing the Prior Informed Consent procedure will also clearly expand, notably in the areas of pesticides and the increasingly ominous domain of the disposal of hazardous wastes. For, as the Register's director has observed: "There are no safe chemicals; there are only safe ways of manufacturing, handling and using them."

DISCUSSION

Question: What is the relationship of IRPTC to the Industry and Environment Programme Activity Centre?
Answer: IRPTC is primarily concerned with hazard assessment risk; we try to help clients decide what chemicals they should and should not buy and thereby prevent the incidence of toxicity. The Industry and Environment Programme Activity Centre works to reduce pollution generally through cleaner production techniques and, through the APELL process, to reduce hazards in emergency situations.

Question: How does IRPTC work with WHO and other United Nations' specialized agencies?
Answer: IRPTC works with both WHO and ILO in the International Programme on Chemical Safety, seeking to evaluate health risks from toxic chemicals. In another joint programme, the Panel of Experts on Environmental Management for

Vector Control, UNEP/IRPTC, FAO and WHO try to deal with the toxic effects of pesticides.

Chapter Fifty-One

The Eastern Africa Environmental Network: an example of networking among developing countries

NATHANIEL ARAP CHUMO
IUCN Regional Committee for East Africa

The Eastern Africa Environmental Network (EAEN) is a regional environmental forum that facilitates the sharing of knowledge and exchange of information by representatives of government agencies, NGOs and individuals. It is a 'market place' for people who care about planet Earth and its inhabitants. These people include environmental policy-makers and development practitioners.

Founded in September 1990, under the impetus of IUCN (Commission on Education), UNEP and NEEN, EAEN used the New England Environmental Network (NEEN), the largest environmental network in the United States, as its model. The Network's headquarters are located in the new building of the East African Wildlife Society in Nairobi, Museum Hill. Run by seven environmental leaders from the region, EAEN activities cover seven eastern African countries: Ethiopia, Djibouti, Kenya, Somalia, Sudan, Tanzania and Uganda. The Network maintains contact with more than one hundred environmental organizations within and outside the region.

There is no doubt that networking in environmental conservation is vital today. It enables organizations to share and coordinate conservation ideas in the course of planning and implementing programmes. Networking discourages the duplication of efforts and helps people realize that they are not working in isolation. It therefore serves as a cohesive force in dealing with environmental problems at national, regional and global levels.

Consequently, EAEN has the following objectives:

1. to stimulate regional communication on issues of environmental concern;

2. to facilitate access to and exchange of information and expertise on environmental issues;

3. to support eastern African conservation initiatives that promote awareness of and commitment to natural resources conservation;

4. to create a forum for discussion on development and environment.

ORIGIN

Although EAEN was founded in September 1990, the impulse to create a network in the region can be traced to an international conference held in Canada in 1971, which was organized by the International Youth Federation (IYF) and its Canadian counterpart to prepare for the Stockholm Conference the following year. Among the recommendations of the IYF-sponsored Canadian Conference was the development of an African youth course in environmental conservation.

A project outline was drawn up by the International Union for the Conservation of Nature and Natural Resources (IUCN) in 1971 and was subsequently developed by a coalition of Kenyan environmental leaders and the UNESCO Regional Office for Science and Technology in Nairobi. An Eastern African Youth Course in Environmental Conservation was held in Nairobi from 4–16 March 1974 and attended by representatives of eight English-speaking eastern African countries. At the end of the course the participants recommended that an interim secretariat be set up in Nairobi to prepare a quarterly bulletin named *The Eastern African Environment* to coordinate the activities of member organizations in various countries and that this Secretariat organize training courses on a regular basis.

At this time, elephants were being killed at an alarming rate in Kenya, largely in the wildlife preserve of Tsavo. The Wildlife Clubs of Kenya (WCK) organized a series of public anti-poaching demonstrations in Nairobi and Embu to pressure the government to step up anti-poaching activities. A petition signed by nearly 8000 people was forwarded to President Jomo Kenyatta. The demonstrations received enormous press coverage and served to focus public attention on the importance of the work of NGOs such as WCK. As a result of the demonstration, hunting was banned in Kenya in 1977. Also banned were curio shops offering wildlife products.

In 1979, WCK worked with other NGOs, as well as the Kenya National Environment Secretariat, to organize another African youth course in environmental conservation. Held in July 1979, with funding from IYF, WWF and UNEP, among others, it attracted participants from the entire African continent, most of them, however, from eastern Africa.

At the end of the course, participants recommended the formation of a clearing house to serve the needs of environment groups in the region. Unfortunately neither sufficient funds nor facilitators were available at the time, nor in the years to come, although this need emerged again and again, notably in the course of a number of workshops for Wildlife Club leaders in East Africa.

Finally, in 1987, with the formation of the Eastern African Committee of IUCN's

Commission on Education and Training links were established with the New England Environmental Network, based at Tufts University, Boston, USA, which invited a number of East Africans to attend the annual New England Environmental Conference that year.

With the assistance of the Lincoln Filene Center of Tufts University, UNEP and others, the Eastern Africa Committee of the IUCN Commission on Education and Training organized a workshop in 1989 in Nairobi to review the Committee's activities and work out a three-year action plan. Among its recommendations was the formation of an Eastern African Environmental Network modelled on NEEN.

EAEN ACTIVITIES

Upon its foundation in 1990, EAEN was charged with the following activities:

1. Compiling a directory of organizations/institutions involved in environmental protection and conservation of natural resources in the eastern African Region. This exercise has now been completed. The first edition of the directory has been produced and distributed.

   At about that time UNEP was also discussing with Tufts University the object of starting a joint post-graduate course, and that new relationship also helped the EAEN–NEEN relationship.

2. Producing a quarterly *Newsletter* to feature activities of member organizations/institutions and environmental issues in the region. The first two issues have been produced and distributed.

3. Maintaining contact with member organizations/institutions through exchange of information and supporting environmental projects.

4. Organizing an annual networking conference. The first annual Conference was held in June 1991 at the National Museum of Kenya in Nairobi. Representatives of Ethiopia, Kenya, Sudan, Tanzania, Uganda and the USA as well as several international regional organizations participated. Participants left with enthusiasm and renewed commitment to work for environmental protection. Successful EAEN Conferences were held in 1992 and 1993.

The aim of the EAEN is to bring together people from different walks of life: environmentalists, civil servants, other professionals, and local and international NGOs. The mission is to address the environmental problems facing the region and find practicable and lasting solutions. Participants felt a significant sense of accomplishment that they had taken the first step toward working together for a sustainable future for the region.

KEY POINTS ABOUT NETWORKING

1. Networking is an important tool in environmental protection because it can be used to shape a common political agenda. Agents of change can increase their influence by joining forces.

2. Because there are so many environmental issues confronting governments at all levels, as well as the business community, no single environmental group can keep track of or respond to them all.

3. Networking is an efficient way for environmental groups to share responsibilities for identifying, studying and responding to many environmental issues in the region.

4. Networking also promotes exchange of *resources* among member organizations/institutions. The purpose is to benefit each party through sharing of resources such as funds and available literature.

5. Networking can also be used to train leaders of participating organizations to develop and expand their knowledge and skills. In doing this type of networking, we conduct needs and interest assessment among the leaders to determine the shared subjects they want to address:
   - agree on what procedures should be used to provide resources for the network, among them mailings and conducting different types of training sessions;
   - decide on the most appropriate type of training methods such as seminars, written materials and audio-visual materials.

6. Training networks are increasingly necessary because the ever-changing nature of environmental problems requires that leaders continuously expand their scientific and political knowledge.

STEPS TOWARDS CREATING AND SUSTAINING A SUCCESSFUL NETWORK

1. Developing one's own openness to learning with others.

2. Recruiting and training facilitators so essential to the fostering of networking; leaders must adhere to the logic of participants. Issues and tactics cannot be decided on by the leader alone because the essence of the networking process is teamwork.

3. Sharing binding values as a further bond among network members.

4. Offering tangible benefits such as opportunities to attend seminars and conferences.

5. Fostering early success. This encourages people to support the network, thereby energizing it and empowering its participants with confidence, which, in turn, enhances the credibility of their joint efforts in the eyes of those they hope to influence.

6. Fostering trust. Participants must have confidence in the integrity of the facilitators and other leaders of the network. Leaders must never allow their personal interest and agendas to undermine the interest of the network effort. They must communicate openly and frequently about all issues that are directly or potentially relevant to the network and its members;

7. Avoiding competition among groups. If this is difficult, the nascent competition should be discussed openly, guided by cooperative principles;

8. Maintenance, which is no more and no less than the increasing amount of care and time spent in network meetings, in talking to individual representatives, in sending them appropriate materials, written and otherwise, and in creating a newsletter that truly reflects the needs and interests of the members;

9. Avoiding over-analysis of the concept of networking. Members are far more eager to hear of concrete results. Do not romanticize either networking or its potential achievements.

CONCLUSION

Networking is a tool that helps us to address current and emerging environmental issues as broadly as possible, so their presentation to government authorities and to donors will have strength and credibility. Non-governmental organizations and government agencies work hard to improve the quality of the environment but our resources are scarce. We must learn to share them.

We have all too little time in which to address pressing environmental issues. Given the limitations of our resources and opportunities, environmental education networks provide the most effective, least expensive way of reaching and increasing the number of engaged and committed citizens in the shortest amount of time.

DISCUSSION

Comment: It is not easy to bring together NGOs and government representatives. They are all competing for funds.
Answer: Networking reveals many commonalities of purpose that hitherto competing groups can share. One of the problems of the past has been the lack of communication among concerned parties.

Question: Is an annual conference sufficient to address all the problems facing environmental NGOs and raise funds to support their work?

Answer: Probably not. EAEN has only one conference a year because of its own limited resources. But we look beyond that constraint in the hope of finding forceful objective members who can persuade others to commit resources to a variety of causes.

Bibliography

MAIN

World Resources (1994) *A Guide to the Global Environment, 1994–95*, World Resources Institute and Oxford University Press, Oxford.

IUCN/UNEP/WWF, (1991) *Caring for the Earth, A Strategy for Sustainable Living*, Gland, Switzerland.

Tolba, M. K. (ed.) (1992) *The World Environment 1972–1992, Two Decades of Challenge*, Chapman & Hall & UNEP, London.

Centre for Our Common Future, (1993) *Agenda for Change, A plain language version of Agenda 21*, Geneva.

Kumar, R. and Murck, B. (1992) *On Common Ground*, John Wiley & Sons, Toronto.

OTHER

Bingham, C. (1984) *Resolving Environmental Disputes: A Decade of Experience*, The Conservation Foundation, Washington, DC.

Bourdeau, P. and Green, G. (1989) *Methods for Assessing and Reducing Injury from Chemical Accidents*, SCOPE 40, John Wiley and Sons, Chichester.

Brown, L. R. *et al.* (1991) *State of the World*, World-Watch Institute, Washington, DC.

Brown-Weiss, E. (1988) *In Fairness to Future Generations: International Law, Common Patrimony and Intergenerational Equity*, Transnational Publishers, Dobbs Ferry, NY.

Callicott, J. B. (1989) *In Defence of the Land Ethic: Essays in Environmental Philosophy*, State University of New York Press, Albany, NY.

Dotto, L. (1986) *Planet Earth in Jeopardy*, John Wiley and Sons, Chichester.

Ehrlich, Paul *et al.* (1977) *Ecoscience*, W. H. Freeman & Co., San Francisco.

Elkington, J. and Hales, J. (1989) *The Green Consumer Guide*, Gollancz, London.

Engel, J. R. and Engel, J. G. (eds) (1990) *Ethics of Environment and Development*, Belhaven Press, London.

Forrester, J. W. (1972) *World Dynamics*, Wright-Allen Press, Cambridge, Mass.

Hardin, G. (1968) *The Tragedy of the Commons*.

Meadows, Donella (1989) *Harvesting One Hundred Fold, Key Concepts and Case Studies in Environmental Education*, UNEP, Nairobi.

Myers, Norman (1984) *Gaia: An Atlas of Planet Management*, Doubleday, New York.

UNEP (1988) *Environmental Perspective to the Year 2000 and Beyond*, United Nations Environment Programme, Nairobi.

World Commission on Environment and Development (WCED) (1987) *Our Common Future*, Oxford University Press, Oxford.

REPORTS AND JOURNALS

Blackwell (Oxford) *Integrated Environmental Management Journal.*
IPPF (London) with IUCN and UNFPA, *People & the Planet.*
Salford University (Manchester) *International Journal of Environmental Education and Information.*
UNEP (Nairobi) *Annual Reports of the Executive Director.*
UNEP (Nairobi) *Our Planet.*
UNDP (New York) *Human Development Reports.*
UNESCO (Paris) *Nature and Resources.*
World Bank (Washington) *World Development Reports.*

Index

Index compiled by Geoffrey C. Jones